The

September 1996

British Bus Publishing

The South Midlands Bus Handbook

The South Midlands Bus Handbook is part of the Bus Handbook series that details the fleets of selected bus and coach operators. These handbooks are published by *British Bus Publishing* and cover Scotland, Wales and England north of London, and coach operators for the south. The current list is shown at the end of the book. Together with similar books for southern England buses, published by Capital Transport which we also supply, they provide comprehensive coverage of all the principal operators' fleets in the British Isles. Handbooks for the FirstBus Group and Stagecoach are also published annually, and a book for the Cowie Group is expected in 1997.

The operators included in this edition are those who are based, and provide stage and express services in English Heartland comprising the midlands counties of Gloucestershire, Oxfordshire, Warwickshire, Herefordshire and Worcestershire. Also included are a number of operators who provide significant coaching activities.

Quality photographs for inclusion in the series are welcome and a fee is payable. The publishers unfortunately cannot accept responsibility for any loss and request you show your name on each picture or slide. Details of changes to fleet information are also welcome.

To keep the fleet information up to date we recommend the Ian Allan publication, *Buses* published monthly, or for more detailed information, the PSV Circle monthly news sheets.

The writer and publisher would be glad to hear from readers should any information be available which corrects or enhances that given in this publication.

Series Editor: Bill Potter
Principal Editors for *The South Wales Bus Handbook*:
Bill Potter and David Donati

Acknowledgements:
We are grateful to Andy Chown, Steve Curl, Robert Edworthy, Mike Fowler, Tony Hunter, Mark Jameson, Martin Perry, the PSV Circle and the operating companies for their assistance in the compilation of this book.

The cover photograph is by Tony Wilson

Contents correct to September 1996

ISBN 1 897990 15 4
Published by *British Bus Publishing Ltd*
The Vyne, 16 St Margarets Drive, Wellington,
Telford, Shropshire, TF1 3PH

CONTENTS

ALEXCARS

Alexcars Ltd, Unit 11, Love Lane Trading Estate, Cirencester, GL7 1YG

HIL7772	Leyland Leopard PSU3E/4R	Willowbrook Warrior (1991)	B48F	1980	Ex Kinch, Barrow-on-Soar, 1996
ACH53A	Bedford YMQ	Plaxton Supreme V Express	C45F	1982	Ex The Delaine, 1989
ODW459	MAN SR280	MAN	C51F	1982	Ex Burton, Finchley, 1989
516ACH	MAN SR280	MAN	C49FT	1983	Ex MAN demonstrator, 1983
ACH69A	Bedford YNV Venturer	Plaxton Paramount 3200 II	C53F	1985	Ex Rambler Coaches, Hastings, 1992
ACH84A	Bedford YNV Venturer	Duple Laser 2	C53F	1985	Ex Smith's, Cymmer, 1992
ACH80A	Bedford YNV Venturer	Duple Laser 2	C55F	1986	Ex Ardenvale, Knowle, 1994
C899REG	Bedford YNV Venturer	Duple 320	C53FT	1986	Ex Swanbrook, Cheltenham, 1990
D27UCW	Freight Rover Sherpa	Videofit	M16	1989	Ex Van, 1989
G140GOJ	Leyland-DAF 400	Leyland-DAF	M16	1990	Ex Dave's, Cirencester, 1994
H39UNH	MAN 10.180	Jonckheere Deauville P599	C37F	1990	Ex Taylor's of Sutton Scotney, 1995
K2SUP	Iveco 315.8.17	Lorraine	C26F	1992	Ex Supreme, Hadleigh, 1994
K201GRW	Dennis Javelin 12SDA2101	Caetano Algarve II	C53F	1993	Ex Supreme, Coventry, 1996

Previous Registrations:

516ACH	UAM932Y	ACH80A	C530UUT	HIL7772	TPT25V
ACH53A	DFE503X	ACH84A	C446HLG	ODW449	RAM72Y
ACH69A	B888PDY				

Livery: Two tone blue

Alec Hibberd started operating coaches in 1949. A weekday service from Cirencester to Tetbury is operated usually employing the Willowbrook Warrier-bodied Leopard. Representing the Alexcars fleet is K2SUP, an Iveco 315 with Lorraine bodywork, a rare combination in Britain. Iveco was formed in 1975 following an agreement between Fiat SpA of Turin and Klöckner-Humboldt-Deutz of Germany, who subsequently withdrew. In 1986 a marketing agreement was signed with Ford of Britain that allowed a joint venture, Iveco Ford Truck Ltd to sell the truck ranges of both manufacturers in the UK. The current PCVs from Iveco are the 380 Euroclass and the UR490 TurboCity. *Robert Edworthy*

ANDY JAMES

A R James, Priory Ind. Est., London Road, Tetbury, GL8 8HZ

	YFR491R	Leyland Leopard PSU3E/4R	Duple Dominant I	C49F	1977	Ex Marshall, Sutton-on-Trent, 1994
	UMR197T	Leyland Fleetline FE30AGR	Eastern Coach Works	H43/31F	1978	Ex Thamesdown, 1994
	UMR198T	Leyland Fleetline FE30AGR	Eastern Coach Works	H43/31F	1978	Ex Thamesdown, 1994
u	FNP98W	Leyland Leopard PSU5D/4R	Duple Dominant III	C57F	1980	Ex Go Whittle, Kidderminster, 1996
	UWY83X	Leyland Leopard PSU3F/4R	Duple Dominant IV Express	C49F	1981	Ex Harrogate & District, 1994
	UWY84X	Leyland Leopard PSU3F/4R	Duple Dominant IV Express	C49F	1981	Ex Harrogate & District, 1994
	GTP95X	Dennis Lancet SD504	Wadham Stringer Vanguard	DP33F	1982	Ex Thamesdown, 1996
u	GTP97X	Dennis Lancet SD504	Wadham Stringer Vanguard	B35F	1982	Ex Thamesdown, 1996
u	AJF68A	Bova FHD12.280	Bova Futura	C49F	1984	Ex Landtourers, Farnham, 1996
	C438SJU	Ford Transit 190	Robin Hood	B16F	1985	Ex Hallam, Newthorpe, 1995
u	D	Leyland Tiger TRCTL11/3R	Wadham Stringer Vanguard	DP F	19	Ex MoD, 1996
	D67OVP	Freight Rover Sherpa	Robin Hood	B21F	1986	Ex Tourist Coaches, Figheldean, 1993
	F57YBO	Leyland Tiger TRCL10/3ARZM	Duple 340	C51F	1989	Ex Bebb, Llantwit Fardre, 1995
	K30ARJ	Renault Trafic	Devon Conversions	M11	1992	Ex MoD, 1996
	M300ARJ	Mercedes-Benz 709D	Autobus Classique	B25F	1994	
	M30ARJ	Mercedes-Benz 711D	Autobus Classique	C24F	1994	
	N30ARJ	Mercedes-Benz 814D	Plaxton Beaver	C33F	1995	
	N3ARJ	Bova FLC12.280	Bova Futura Club	C53F	1995	

Previous Registrations:

AJF68A A796KEP, TSU601 FNP98W KUX222W, URH341

Livery Cream and yellow

Andy James turns out a smart white and yellow liveried fleet. Refurbished to exhibition standards is YFR491R, a Leyland Leopard with Duple Dominant I bodywork built for National Bus and allocated new to Ribble. The Duple Dominant I body was built only with grant-door treatment. The operator is also involved in dealing and undertaking major refurbishments, so some vehicles listed may not enter service. A number of services are operated in the Cirencester/Chippenham/Malmesbury and Tetbury area. *Phillip Stephenson*

ASTONS

Astons of Kempsey (Coaches) Ltd; TW&SD Halford; Clerkenleap Farm, Kempsey, Worcestershire WR5

HJI531	Volvo B58-56	Plaxton Elite III	C53F	1972	Ex Bennetts, Gloucester, 1989
DSU114	Volvo B58-56	Plaxton P'mount 3200 (1986)	C53F	1973	Ex Pullman, Penclawdd, 1992
GGD664T	Volvo B58-61	Plaxton Supreme IV	C57F	1979	Ex Pykett, Kilburn, 1989
NIA5055	Volvo B58-61	Plaxton Supreme IV	C57F	1980	Ex Thandi, Smethwick, 1989
A676EYJ	Mercedes-Benz L508D	Mercedes-Benz	M14	1984	Ex ? Hospital, 1992
IIB145	Volvo B10M-61	Jonckheere Jubilee P50	C49F	1985	
HYY3	Volvo B10M-61	Jonckheere Jubilee P50	C57F	1985	Ex Buddens, Woodfalls, 1988
F112TEE	Mercedes-Benz 507D	Coachcraft	M10L	1989	Ex B&D Richards, Cimla, 1993
G490PNF	Mercedes-Benz 508D	Made-to-Measure	C20F	1989	Ex Berry, Stockton, 1993
G577RNC	Peugeot-Talbot Express	Made-to-Measure	M8	1989	
H726UKY	Ford Transit VE6	Advanced Vehicle Bodies	M14	1991	Ex Kenning Car Hire, 1993
J336UHP	Peugeot-Talbot Express	Talbot	M8	1992	
GSC858T	Leyland Fleetline FE30AGR	Eastern Coach Works	H43/33F	1978	Ex Malvernian Tours, 1995
ULS670T	Leyland Fleetline FE30AGR	Eastern Coach Works	H43/32F	1979	Ex Malvernian Tours, 1995
MCT226X	Volvo B58-56	Duple Dominant II Express	C57F	1982	Ex Brylaine, Boston, 1995
RDU4	Volvo B10M-61	Duple 340	C57F	1987	Ex Country Lion, Northampton, 1994
A1FRP	Volvo B10M-61	Van Hool Alizée	C53F	1988	Ex Leon's, Stafford, 1995
E26JBD	Ford Transit VE6	Ford	M12	1988	Ex Northamptonshire CC, 1995
E245CGA	DAF SB2305DHS585	Duple 340	C53FT	1988	Ex Starline Coaches, Woolton, 1996
F752JLG	Ford Transit VE6	Steedrive	M16	1989	Ex Self Drive Hire, 1995
F944KTA	Volkswagen Transporter	Devon Conversions	M11	1989	Ex Dave Parry, Cheslyn Hay, 1994
G134GOL	Iveco Daily 40.8	Carlyle Dailybus	B21F	1990	Ex ?, 1996

Astons livery is unusual in that the name is applied like ribbons that trail around the vehicle to suit the styling. Pictured in Parliament Square, London, is L360YNR, one of three Dennis Javelins in the fleet. A new depot, replacing the small premises in the centre of Kempsey opened late in 1996 and is situated nearer to Worcester. *Colin Lloyd*

Photographed on service in Worcester during May 1996 is Astons K544OGA, a Mercedes-Benz 711D with bodywork from Dormobile to the Routemaker design. The vehicle can be found on several tendered school services for Hereford & Worcester County Council. *Richard Godfrey*

24PAE	DAF SB2305DHS585	Duple 340	C53FT	1990	Ex BAA, Gatwick, 1996
G965PRC	Ford Transit VE6	Ford	M8	1990	Ex private owner
3698E	Volvo B10M-56	Jonckheere Deauville P599	C51F	1991	Ex Harry Shaw, 1994
K544OGA	Mercedes-Benz 711D	Dormobile Routemaker	B29F	1992	Ex Nuttall, Penwortham, 1995
L360YNR	Dennis Javelin 12SDA2131	Plaxton Premiere 320	C53FT	1994	
L230BUT	Dennis Javelin 12SDA2131	Plaxton Premiere 320	C53FT	1994	
N760NAY	Dennis Javelin 12SDA2136	Marcopolo Explorer	C51FT	1995	
N914DWJ	Scania K113CRB	Van Hool Alizée	C49FT	1996	
N915DWJ	Scania K113CRB	Irizar Century 12.35	C49FT	1996	
P63GHE	Scania K113CRB	Van Hool Alizée	C49FT	1996	

Previous Registrations:

24PAE	G987KJX	HJI531	XWX797L
3698E	H170NHP, H5URE	HYY3	C405LRP
A1FRP	E275HRY	IIB145	B968CWP, RDU4, B262MAB
DSU114	JWO946L	NIA5055	GJU855V
E245CGA	E665KCX, A9KRT	RDU4	D36ENH, A8CLN, D146ENV
E742CDS	E749JAY, A4KRT		

Livery: White, grey and brown; dark grey (coaches); white (Eurolines) N914/5DWJ

BARRY'S COACHES

R B Talbott; Meon Valley Coaches Ltd, Parker's Lane, Stow Road, Moreton-in-the-Marsh, Gloucestershire, GL56 0DP

JOV766P	Ailsa B55-10	Alexander AV	H44/35F	1976	Ex London Buses, 1993
MED396P	Ford R1114	Duple Dominant	C53F	1976	Ex Smith's, Wigan, 1980
IIL1237	Volvo B58-61	Plaxton Supreme III	C57F	1978	Ex Catteralls, Southam, 1994
CFX320T	Ford R1014	Plaxton Supreme IV	C41F	1979	Ex Mullover, Bedford, 1986
EPM137V	AEC Reliance 6U2R	Plaxton Supreme IV Express	C53F	1979	Ex London Country, 1986
KAD351V	Leyland Leopard PSU5C/4R	Plaxton Supreme IV	C57F	1980	Ex Southend, 1990
JDB950V	Ford R1114	Plaxton Supreme IV	C53F	1980	Ex Wallington, Great Rollright, 1993
LFH675V	Ford R1114	Duple Dominant II Express	C53F	1980	
TND431X	Volvo B10M-61	Plaxton Supreme V	C53F	1982	Ex Smith Shearings, 1990
CIL3526	DAF MB200DKTL600	Plaxton Paramount 3200	C51F	1983	Ex Smith Shearings, 1988
GIL4267	Volvo B10M-61	Caetano Algarve	C51FT	1986	Ex Park's, 1988
E346AAM	Toyota Coaster HB31R	Caetano Optimo	C21F	1988	Ex Ellison, Ashton Keynes, 1993
F700PAY	Mercedes-Benz 0303	Mercedes-Benz	C53F	1988	Ex Redwing, Camberwell, 1996
F254MGB	Volvo B10M-61	Van Hool Alizée	C49FT	1989	Ex Park's, 1995
L111RBT	MAN 18-370	Berkhof Excellence 1000	C51FT	1994	

Previous Registrations:

CIL3526	ANA453Y	GIL4267	C703KDS
F254MGB	F767ENE, LSK508	IIL1237	XEH1S, 9530RU, ARF150S

Livery: Green and gold; white, maroon and purple (modern coaches)

Mr Talbott entered the coach business 45 years ago in 1951 and took over the established Meon Valley Coaches business in 1975. Operations extend into Warwickshire and Oxfordshire with several local and market-day services provided. Photographed on a homeward journey from school, TND431X of Barry's Coaches is passing through Evesham for the village of Shipston. New to Shearings it was one of the first of many B10M to operate on their tours and is one of two from the chassis supplier to operate with Barry's. *Richard Godfrey.*

K W BEARD LTD

K W Beard Ltd, Valley Road, Cinderford, GL14 2PD

	NVJ150M	Bedford YRT	Plaxton Elite III	C53F	1973	Ex Yeoman's, Canon Pyon, 1980
	NNW119P	Leyland Leopard PSU3C/4R	Duple Dominant	C53F	1976	Ex Wallace Arnold, 1983
w	MHB851P	Leyland Leopard PSU3C/4R	Plaxton Supreme III Express	C53F	1976	Ex Hill's of Tredegar, 1991
	VYL851S	Bedford YMT	Van Hool McArdle 300	C53F	1978	Ex Anglo-French Cs, Chislehurst, 1986
	YPL72T	AEC Reliance 6U2R	Duple Dominant II Express	C53F	1979	Ex London Country, 1985
	WDD17X	Bedford YNT	Plaxton Supreme VI Express	C53F	1982	
	LIL9968	Leyland Tiger TRCTL11/3R	Plaxton Viewmaster IV	C49F	1982	Ex Karvien, Walsall, 1994
	OHE275X	Leyland Tiger TRCTL11/3R	Duple Dominant IV	C53F	1982	Ex West Riding, 1987
	PJI7755	Bova EL28/581	Duple Calypso	C53F	1984	Ex John Speck Coaches, Ledbury, 1995
	B502UNB	Leyland Tiger TRCTL11/3RZ	Plaxton Paramount 3500 II	C53F	1985	Ex Shearings, 1990
	E318UUB	Volvo B10M-61	Plaxton Paramount 3500 III	C53F	1987	Ex Ashton, St Helens, 1995
	F715RDG	Freight Rover Sherpa	Crystals	M16	1988	
	F167UDG	Leyland Tiger TRCTL11/3RZ	Plaxton Paramount 3200 III	C53F	1989	

Previous Registrations:

PJI7755 A332PFJ

LIL9968 EFE261X, 5447FH, FTL883X, ORJ701, XBC350X

Livery White and two-tone blue

Introduced in April 1983, the Duple Calypso was a joint venture between Autobus-fabrek Bova, the Dutch vehicle manufacturer and Duple, intended as an alternative to the Europa for operators who wanted a British-built integral coach. It shares many body parts with other Duple models. Seen in the livery of K W Beard Ltd is PJI7755. The operators depot was previously employed by the Moseley Group as a PSV dealership and as the Welsh-based Red & White's outstation with three vehicles on the premises. *Richard Edworthy*

BENNETTS

R A, D & P Bennett and P A Lane, Eastern Avenue, Gloucester, GL4 7BU

WWY125S	Bristol VRT/SL3/6LXB	Eastern Coach Works	H43/31F	1978	Ex Red Bus Services, Aylesbeare, 1994
HAX331W	Ford R1114	Plaxton Supreme IV	C53F	1980	Ex Prior Park College, Bath, 1990
RUA451W	Bristol VRT/SL3/6LXB	Eastern Coach Works	H43/31F	1980	Ex Yorkshire Buses, 1994
RUA452W	Bristol VRT/SL3/6LXB	Eastern Coach Works	H43/31F	1980	Ex Yorkshire Buses, 1994
RUA457W	Bristol VRT/SL3/6LXB	Eastern Coach Works	H43/31F	1981	Ex Yorkshire Buses, 1993
RUA458W	Bristol VRT/SL3/6LXB	Eastern Coach Works	H43/31F	1981	Ex Yorkshire Buses, 1993
RUA460W	Bristol VRT/SL3/6LXB	Eastern Coach Works	H43/31F	1981	Ex Yorkshire Buses, 1994
LPY459W	Leyland Leopard PSU3E/4R	Duple Dominant	B55F	1981	Ex United, 1993
XAD174X	Ford R1114	Duple Dominant Express	C53F	1982	
TVH137X	Ford R1014	Plaxton Supreme V	C35F	1982	Ex Select, Hartlepool, 1994
PBO11Y	Ford R1114	Plaxton Supreme V	C53F	1982	Ex Capitol Coaches, Cwmbran, 1989
PJI7230	DAF MB200DKFL600	Duple Laser	C57F	1984	Ex Nichols, Carlton, 1996
B472ENT	Leyland Tiger TRCTL11/2R	Duple Dominant	B55F	1984	Ex Thomas Bros., Llandeilo, 1992
B368YDE	DAF SB2300DHS585	Plaxton Paramount 3200	C55F	1985	Ex Silcox, Pembroke Dock, 1987
D881BDF	DAF SB2300DHTD585	Plaxton Paramount 3200 II	C55F	1986	
D290XCX	DAF SB2300DHS585	Plaxton Paramount 3200 III	C53F	1987	Ex Smith's Coaches, Alcester, 1988
F318EWF	DAF SB220LC550	Optare Delta	DP49F	1988	Ex Wall's, Northenden, 1995
F642KCX	DAF MB230LB615	Plaxton Paramount 3500 III	C53F	1988	Ex Smith's Coaches, Alcester, 1988
F643OHD	DAF SB2305DHS585	Van Hool Alizée	C51FT	1989	Ex Smith's Coaches, Alcester, 1990
F607JSS	DAF SB2305DHS585	Caetano Algarve	C53F	1989	Ex Whyte & Urquhart, Newmachar, 1994
G905WAY	DAF SB2305DHS585	Caetano Algarve 2	C53F	1989	Ex Bourne & Woodhall, Laindon, 1991
G382RCW	DAF SB2305DHS585	Van Hool Alizée	C53F	1990	Ex Armchair, Brentford, 1994
G975KJX	DAF SB2305DHS585	Van Hool Alizée	C51FT	1990	Ex Armchair, Brentford, 1994
H11PSV	DAF SB2305DHS585	Plaxton Paramount 3200 III	C53F	1991	
J21GCX	DAF SB2305DHS585	Plaxton Paramount 3200 III	C53F	1991	Ex Yorkshire European, Harrogate, 1992
K2BCC	DAF MB230LT615	Plaxton Paramount 3500 III	C53F	1992	
L463RDN	DAF SB2700HS585	Van Hool Alizée	C51F	1994	Ex ?, 1995

Previous Registrations:

B368YDE	817FKH	L493RDN	?
G382RCW	G973KJX, BIB5491	PJI7230	A335KLK
HAX331W	HAX335W	TVH137X	TVH137X, PSV111

Livery Blue, grey and orange

One of two single-deck buses with Bennetts, LPY459W is a Leyland Leopard with Duple Dominant bus bodywork. Liveried for school duties, it was photographed while heading for Gloucester. *Robert Edworthy*

Prepared for both bus and coach work at Bennetts is F318EWF, a DAF SB220 with Optare Delta dual-purpose bodywork. It was seen on Gloucestershire service 534 to Newent which has been run since the mid 1970s. Previously the vehicle operated the infamous Wilmslow Road services in Manchester for Wall's. *Robert Edworthy*

Double-deck buses are used by Bennetts for school contracts with six Bristol VRs now operated. Most have come from the same batch at Yorkshire Buses and one of these, RUA460W, is seen in the centre of Gloucester. Park & Ride services were started by Bennetts in 1992 and 1993. *Les Peters*

BOOMERANG BUS CO

R C & N J Warner Ltd; Boomerang Bus Co Ltd, Oldbury Buildings, Northway Lane, Tewkesbury, Gloucestershire, GL20 8JG

RL2727	Thornycroft A1	Mumford	Ch20	1926	Ex preservation
ABH358	Leyland Cub KP3	Duple	C20F	1932	Ex preservation
LIL3065	Leyland Leopard PSU3E/4R	Plaxton Supreme IV Express	C53F	1980	Ex Vanguard, 1993
CTX397V	Bristol VRT/SL3/6LXB	Alexander AL	H44/31F	1980	Ex Cardiff Bus, 1994
LIL3066	Leyland Leopard PSU3F/5R	Plaxton Supreme IV Express	C49F	1980	Ex Nelson, Thornhill, 1994
SGS499W	Leyland Tiger TRCTL11/3R	Plaxton Supreme IV	C50F	1981	Ex S&P Coaches, Southend on Sea, 1996
6017WF	Bedford YMQ	Duple Dominant IV Express	C35F	1981	Ex Oakley Coaches, 1993
5904WF	Leyland Tiger TRCTL11/3R	Duple Goldliner IV	C50F	1982	Ex Leyland Motors, Leyland, 1984
8921WF	Leyland Tiger TRCTL11/3R	Van Hool Alizée	C49FT	1982	Ex Leyland Motors, Leyland, 1984
LIL9270	Leyland Tiger TRCTL11/3R	Plaxton Paramount 3200	C53F	1983	Ex Rider Group, 1995
E62SUH	Volkswagen LT55	Optare City Pacer	B25F	1988	Ex Rees & Williams, Morriston, 1994
A13WMS	Renault Master T35D	Coachcraft	M13	1988	Ex Ashfield Travel, Kirkby in Ashfield, 1994
E164TWO	Freight Rover Sherpa (Isuzu)	Carlyle Citybus 2	B20F	1988	Ex HB Transport, Faversham, 1993
E170TWO	Freight Rover Sherpa (Isuzu)	Carlyle Citybus 2	B20F	1988	Ex HB Transport, Faversham, 1993
E175TWO	Freight Rover Sherpa (Isuzu)	Carlyle Citybus 2	B20F	1988	Ex HB Transport, Faversham, 1993
F905YWY	Mercedes-Benz 811D	Optare StarRider	B26F	1988	Ex Cowie South London, 1996
F915YWY	Mercedes-Benz 811D	Optare StarRider	B26F	1988	Ex Cowie South London, 1996
4529WF	Freight Rover Sherpa (Isuzu)	Carlyle Citybus 2	B20F	1989	Ex Carlyle, Birmingham, 1991
LIL9267	Mercedes-Benz 811D	Phoenix	B31F	1990	Ex Solent Blue Line, 1996
LIL9268	Mercedes-Benz 811D	Phoenix	B31F	1990	Ex Solent Blue Line, 1996

Previous Registrations:

4529WF	F418BOP	A13WMS	E710YWE	LIL9267	G209YDL
5904WF	THG852X, 3081WF	C745RKB	C555LMW, AIA5505	LIL9268	G210YDL
6017WF	NJT120W	LIL3065	KUB545V	LIL9270	EWW947Y
8921WF	WBV541Y, 6449WF	LIL3066	BVA787V		

Previously with Cardiff Bus, CTX397V now sports a livery of yellow and black for school contract work with Warners and the Boomerang Bus Company. This batch of Bristol VRs carry Alexander AL bodywork rather than the more common Eastern Coach Works product. Warners can trace their roots to the 1920s although the present business was founded in 1988. Minibuses are employed on Tewkesbury and Worcester town services.
Robert Edworthy

BROMYARD OMNIBUS CO

M Perry, Streamhall Garage, Linton Trading Estate, Bromyard, Herefordshire, HR7 4QL

w	JDK911P	Bristol RESL6L	East Lancashire	DP42F	1975	Ex Devon Services, 1993
50	ORS86R	Leyland Leopard PSU4E/4R	Alexander AY	DP45F	1977	Ex Grampian, 1992
98	TUP432R	Bedford YRT	Plaxton Derwent	DP53F	1976	Ex Cave, Shirley, 1996
97	ARB528T	Bedford YMT	Plaxton Supreme III Express	DP49F	1978	Ex East Midland, 1996
70	OHV188Y	Ford R1114	Wadham Stringer Vanguard	B32F	1982	Ex Croydon Coaches, Newport, 1995
74	XNK201X	Ford R1014	Plaxton Bustler	B47F	1981	Ex Abbey Garage, Newburgh, 1994
75	6795FN	Ford R1014	Plaxton Bustler	B47F	1981	Ex Flagfinders, Braintree, 1994
79	C531TJF	Ford Transit	Rootes	B16F	1986	Ex Midland Fox, 1995
99	D917GRU	Bedford YNT	Plaxton Derwent	B53F	1987	Ex Western Scottish, 1996
81	D969PJA	Renault-Dodge S56	Northern Counties	B20F	1987	Ex Little Red Bus, Smethwick, 1995
85	D166RAK	Renault-Dodge S56	Reeve Burgess	B25F	1987	Ex Mainline, 1996

Livery: Red and cream

Previous Registrations:
6795FN

One of Martin Perry's Bromyard Omnibus Company pair of Ford R1014s with Plaxton Bustler service bus bodies is 6795FN. It is seen in Bromyard operating town service 400, during July 1996. The Bustler was introduced in 1981 with many of the early examples heading for Heathrow where they were employed on duties between terminals. *Martin Perry*

CARTERTON COACHES

Carterton Coaches (Witney) Ltd, 32 Corn Street, Witney, OX8 7BN

Depot: Downs Road, Witney

461BDL	AEC Reliance 6U3ZR	Duple Dominant	C53F	1976	Ex Bampton Travel, Witney, 1996
NAL53P	Leyland Fleetline FE30AGR	Alexander AD	H44/34F	1976	Ex Swanbrook, Cheltenham, 1993
NDD113W	Bedford YMT	Plaxton Supreme IV Express	C53F	1980	Ex Swanbrook, Cheltenham, 1993
A463JJF	Fiat 60F10	Caetano Beja	C18F	1984	Ex McDougall, Hoddesdon, 1994
RBT172M	AEC Reliance 6MU4R	Plaxton Elite III	C53F	1974	Ex Common, Witney, 1995
YTD384N	Bedford YRT	Plaxton Elite III Express	C53F	1974	Ex SP Private Hire, Eynsham, 1995
HFG451T	Ford R1014	Duple Dominant II	C35F	1979	Ex Golder, Stratford, 1996
XSU912	MCW Metroliner HR131/2	MCW	C49FT	1984	Ex Stagecoach South (East Kent), 1995
LDZ3145	MCW Metroliner HR131/6	MCW	C49FT	1985	Ex Stagecoach South (East Kent), 1995

Previous Registrations:

461BDL	MED410P	LDZ3145	B245JVA
HFG451T	APH519T, 9925AP	XSU912	B192JVA

Livery: White, yellow, orange and red

Recent withdrawals from Carterton Coaches include this Plaxton-bodied Scania, GDZ8449 photographed at the Swindon end of Carterton's Oxford-Swindon service. Regular performers on the service are a pair of MCW coaches previously with Stagecoach South where they were allocated to the East Kent operation to which they had been new. *Andrew Jarosz*

CATTERALLS

Portrest Ltd, 14B Banbury Road, Southam, Warwickshire, CV33 OEA

Depot: Long Itchington

THX634S	Leyland Fleetline FE30ALR	Park Royal	H44/27D	1978	Ex Filer, Ilfracombe, 1995
BVP802V	Leyland Leopard PSU3E/4R	Willowbrook 003	C53F	1980	Ex Travel de Courcey, Coventry, 1994
HIL8516	Volvo B10M-61	Van Hool Alizée	C49FT	1981	Ex Travel De Courcey, Coventry, 1995
GOI7376	MAN SR280	MAN	C49FT	1981	Ex ?, 1996
TND115X	Volvo B58-61	Duple Dominant IV	C53F	1982	Ex Bryan A Garratt, Leicester, 1994
SWV804	Auwaerter Neoplan N216H	Auwaerter Jetliner	C14FT	1982	Ex Aston's, Kempsey, 1995
A21PFR	Bedford YNT	Duple Laser	C53F	1984	Ex Hill, Congleton, 1994
4849RU	Scania K112TRS	Berkhof Eclipse	CH36/10CT	1986	Ex Astons, Kempsey, 1992
LIL2050	DAF MB230LB615	Caetano Algarve	C49FT	1988	
L349MKU	Volvo B10M-62	Plaxton Excalibur	C49FT	1994	Ex MacMillan, Stratford, 1995

Previous Registrations:

4849RU	D290RHK, HOL5W, D438BGX, RDU4, D93DWP
GOI7376	LHR830X
HIL8516	STT607X, XCD108, YDP694X
LIL2050	E300ONH
SWV804	KVJ1Y, 267ALP

Livery: White, yellow and orange

The newest coach with Catteralls is L349MKU, a Volvo B10M with Plaxton Excalibur bodywork. The high- specification luxury coach is used and lettered for the operator's European Holidays. Compared with the Première body the Excalibur has a sloping windscreen and generally features a higher level of equipment. *Robert Edworthy*

CASTLEWAYS

Castleways (Winchcombe) Ltd, Castle House, Greet Road, Winchcombe, Gloucestershire, GL54 5PH

EDG250L	Leyland Leopard PSU3B/4R	Plaxton Elite III Express	C51F	1972	
FAD708T	Bedford YMT	Plaxton Supreme IV Express	C49F	1979	
LFH719V	Leyland Leopard PSU3E/4R	Plaxton Supreme IV Express	C49F	1980	
LFH720V	Leyland Leopard PSU3E/4R	Plaxton Supreme IV Express	C49F	1980	
TJF757	Kässbohrer Setra S215HR	Kässbohrer Rational	C53F	1988	
G82BHP	Peugeot Talbot Pullman	Talbot	C20F	1990	Ex Sapwell, Ashton, 1993
86JBF	Van Hool T815	Van Hool Alicron	C49FT	1990	
H383HFH	Mercedes-Benz 814D	Reeve Burgess Beaver	C33F	1991	
J306KFP	Toyota Coaster HDB30R	Caetano Optimo II	C21F	1991	
J362BNW	Mercedes-Benz 811D	Optare StarRider	C29F	1991	
J688MFE	Kässbohrer Setra S215HR	Kässbohrer Rational	C53F	1992	
J689MFE	Kässbohrer Setra S215HR	Kässbohrer Rational	C53F	1992	
M151KDD	Dennis Dart 9.8SDL3054	Plaxton Pointer	DP40F	1995	
N325MFE	Kässbohrer Setra S210HD	Kässbohrer	C35F	1995	

Previous Registrations:

TJF757	E130NFH	86JBF	G901ANR

Livery: Navy blue, grey and gold

Opposite: **One of a pair of Leyland Leopards, LFH719V, is seen here together with another Plaxton-bodied vehicle, M151KDD, a Dennis Dart for Castleways local services, seen here when new.**
Below: **In addition to their local services, Castleways operate high-specification coaches from a base in Winchcombe, Gloucestershire. The fleet includes three Kässbohrer Setra coaches including an S210H, N325MFE, pictured at Cheltenham race course. At one time the Setra type number would indicate the number of rows of seats in a standard vehicle, the S210 having 10 rows and a normal capacity of 39. In 1992 it's successor, the S309 was introduced in left-hand drive form with the first series 3 right-hand drive models expected this year.** *David Donati*

CASTLEWAYS
CASTLEWAYS
CHELTENHAM
LFH 719V

559
CASTLEWAYS
FOR QUALITY
CASTLEWAYS
CASTLEWAYS
DENNIS
M151 KDD
PLAXTON
POINTER

CHARLTON SERVICES

NGJ, NHM & PD Holder, The Garage, High Street, Charlton-on-Otmoor, Oxfordshire, OX5 2UQ

GUD708L	Leyland Leopard PSU3B/4R	Plaxton Elite III	C51F	1972	
AAU136A	Leyland Leopard PSU5/4R	Plaxton Elite III	C57F	1973	Ex Lewis, Llanrhystyd, 1992
NWT637P	Leyland Leopard PSU3C/4R	Plaxton Supreme III	C46F	1976	Ex Kidlington Band, 1991
NWT639P	Leyland Leopard PSU3C/4R	Plaxton Supreme III	C53F	1976	Ex Ranger, Farnborough, 1987
VGJ317R	Leyland Leopard PSU5A/4R	Plaxton Supreme III	C55F	1977	Ex Epsom Coaches, 1984
VPP958S	Leyland Leopard PSU5C/4R	Plaxton Supreme III	C57F	1978	Ex Biss Brothers, B Stortford, 1985
EBM440T	Leyland Leopard PSU5C/4R	Plaxton Supreme IV	C50F	1979	Ex Squirrell, Hitcham, 1995
EDF269T	Leyland Leopard PSU5C/4R	Plaxton Supreme IV	C57F	1979	Ex Shamrock & Rambler, 1984
CSU432	Leyland Tiger TRCTL11/3R	Plaxton Paramount 3200	C53F	1982	Ex Raff, Gravesend, 1991
CSU243	Leyland Tiger TRCTL11/3R	Plaxton Paramount 3200	C57F	1983	
OJI3907	Leyland Tiger TRCTL11/3R	Plaxton Paramount 3200	C57F	1983	Ex B J Coaches, Abbey Wood, 1988
GJI7173	Leyland Tiger TRCTL11/3R	Plaxton Paramount 3200	C53F	1983	Ex United, 1989
GAZ8573	Leyland Tiger TRCTL11/3R	Plaxton Paramount 3500	C53F	1983	Ex Lee, Rowlands Gill, 1995
WXI6274	Mercedes-Benz L307D	Reeve Burgess	M12	1984	Ex William Ramsay, Elsrickle, 1988
3103PH	Leyland Tiger TRCTL11/3RH	Plaxton Paramount 3500 II	C46FT	1985	Ex Cheltenham & Gloucester, 1993
LIL2665	Leyland Tiger TRCTL11/3RH	Plaxton Paramount 3500 II	C49FT	1985	Ex Cheltenham & Gloucester, 1993
OXI9100	Leyland Royal Tiger RT	Plaxton Paramount 3500	C55F	1986	Ex Horseman, Reading, 1992
C378PCD	Leyland Tiger TRCTL11/3RH	Plaxton Paramount 3500 II	C51F	1986	Ex Brighton & Hove, 1995
C379PCD	Leyland Tiger TRCTL11/3RH	Plaxton Paramount 3500 II	C51F	1986	Ex Brighton & Hove, 1995

Previous Registrations:

3103PH	B215NDG, 511OHU, B177SFH	GJI7173	AEF29Y
AAU136A	FAU46L	LIL2665	B216NDG, HIL6075, B178SFH
CSU243	FJO603Y	OJI3907	YFG366Y
CSU432	MHR847X	OXI9100	C801FMC
GAZ8573	10RU, GNF980Y	WXI6274	A841UGB

Livery: Two-tone blue

Charlton Services fleet is dominated by Leyland products with models divided between the Leopard and the Tiger. One of the later Leopards is EBM440T, seen here with a Plaxton Supreme IV body as it operates Oxfordshire service 108.
Andrew Jarosz

All of Charlton Services' Leyland Tigers have Plaxton Paramount bodies though both heights are represented and both types shown here for comparison. Pictured in Bath, where the driver and vehicle are resting, is LIL2665, while CSU243 was photographed on a day excursion to Blackpool, still one of the main attractions for coach parties. Note the difference between the original Paramount and the mark II version. *Richard Eversden/Colin Lloyd*

CHAUFFEURS

AW Robinson (Taxis) Ltd, Curdworth Garage, Kingsbury Road, Curdworth, Warwickshire.

GPX546X	Mercedes-Benz L508DG	Robin Hood	C21F	1982	Ex Summerfield, Southampton, 1993
HIL7404	Bova EL26/581	Bova Europa	C47FT	1984	Ex Wood, Bognor Regis, 1991
XXI8563	Scania K112CRS	Jonckheere Jubilee P50	C49FT	1984	Ex The Birmingham Coach Company, 1996
JIL3756	Volvo B10M-61	Plaxton Paramount 3500 II	C49FT	1985	Ex Owen, Oswestry, 1995
SIB8340	DAF SB2300DHS585	Jonckheere Jubilee P50	C28FT	1986	Ex The Birmingham Coach Company, 1995
D291ALR	MAN MT136	G C Smith Whippet	C28F	1986	Ex North Mymms, Potters Bar, 1991
JIL3755	Scania K112CRB	Van Hool Alizée	C55F	1987	Ex Shearings, 1990
JIL8323	Scania K112CRB	Van Hool Alizée	C49FT	1988	Ex Wessex, 1994
JIL3757	Volvo B10M-61	Van Hool Alizée	C53F	1988	Ex Travellers, Hounslow, 1994
F434ENB	Mercedes-Benz 507D	Made-to-Measure	C20F	1988	
F639HVM	Freight Rover Sherpa	Made-to-Measure	M16	1989	
F683UDP	Fiat Ducato	Fiat	M14	1989	Ex Private Owner, 1990
F426RRY	Toyota Coaster HB31R	Caetano Optimo	C18F	1989	Ex Canham, Whittlesey, 1991
G271UFB	Mercedes-Benz 609D	Made-to-Measure	C26F	1989	
G42RGG	Volvo B10M-60	Plaxton Paramount 3500 III	C F	1990	Ex Park's, 1992
H404CJF	Toyota Coaster HB31R	Caetano Optimo	C18F	1990	
H650MBF	Mercedes-Benz 814D	PMT Ami	C30F	1991	
J625KUT	Mercedes-Benz 609D	Whittaker Europa	C21F	1991	Ex Harker, Sudden, 1995
K208SFP	Toyota Coaster HDB30R	Caetano Optimo II	C21F	1992	

Previous Registrations:

HIL7404	436FYM, A451KFP	JIL3757	E265OMT	SIB8340	D321VVV
JIL3755	D359OBA	JIL8323	E664YDT	XXI8563	A67JLW, C753TMY
JIL3756	C259VAJ				

Livery: White and blue

A split-screen Jonckheere Jubilee bodied DAF SB2300 is shown in Chauffeurs livery. Jonckheere are now the Belgian part of the Dutch Berkhof group with the latest product, the Mistral being imported into Britain this year. The picture is taken at the Cheltenham Gold Cup meeting in Spring 1996. *David Donati*

CHELTENHAM & GLOUCESTER

Cheltenham & Gloucester Omnibus Company Ltd
Cheltenham District Traction Company Ltd,
Swindon & District Bus Company Ltd,
3/4 Bath Street, Cheltenham, GL50 1YE
Circle Line, Abbey Road Depot, Hempsted, Gloucester GL2 6HU

Depots : Lansdowne Ind Est, Gloucester Road, Cheltenham; Cirencester; London Road, Gloucester; London Road, Stroud and Eastcott Road, Swindon. **Outstation:** Love Lane, Gloucester.

101-105		Leyland Olympian ONLXB/2RZ		Alexander RL		H51/36F	1990		
101	G101AAD	**102**	G102AAD	**103**	G103AAD	**104**	G104AAD	**105**	G105AAD

106-111		Leyland Titan TNLXB/1RF		Park Royal		H47/26F	1979-80 Ex Thames Transit, 1990		
106	GNF6V	**108**	GNF8V	**109**	GNF9V	**110**	GNF10V	**111**	GNF11V

112-124		Leyland Olympian ONLXB/1R		Roe		H47/29F	1982-83 Ex Bristol, 1983 113 ex Yorkshire Rider, 1987		
112	JHU899X	**115**	LWS33Y	**118**	LWS36Y	**121**	LWS39Y	**123**	LWS41Y
113	UWW7X	**116**	LWS34Y	**119**	LWS37Y	**122**	LWS40Y	**124**	NTC132Y
114	JHU912X	**117**	LWS35Y	**120**	LWS38Y				

201	JOU160P	Bristol VRT/SL3/501(6LXB)	Eastern Coach Works	H43/28F	1975	Ex Bristol, 1983
202	MUA872P	Bristol VRT/SL3/6LX	Eastern Coach Works	H43/31F	1975	Ex Bristol, 1983
204	MOU739R	Bristol VRT/SL3/6LXB	Eastern Coach Works	H43/28F	1976	Ex Bristol, 1983
205	HNU670R	Bristol VRT/SL3/6LXB	Eastern Coach Works	H43/28F	1976	Ex Bristol, 1983
209	NWS289R	Bristol VRT/SL3/6LXB	Eastern Coach Works	H43/28F	1976	Ex Bristol, 1983
213	REU311S	Bristol VRT/SL3/6LXB	Eastern Coach Works	H43/28F	1977	Ex Bristol, 1983
215	XDV602S	Bristol VRT/SL3/6LXB	Eastern Coach Works	H43/29F	1978	Ex Devon General, 1987
216	XDV606S	Bristol VRT/SL3/6LXB	Eastern Coach Works	H43/28F	1978	Ex Devon General, 1987
218	VOD596S	Bristol VRT/SL3/6LXB	Eastern Coach Works	H43/28F	1978	Ex Devon General, 1987
219	VOD597S	Bristol VRT/SL3/6LXB	Eastern Coach Works	H43/28F	1978	Ex Devon General, 1987
222	TWS906T	Bristol VRT/SL3/6LXB	Eastern Coach Works	H43/28F	1979	Ex Bristol, 1983
223	TWS913T	Bristol VRT/SL3/6LXB	Eastern Coach Works	H43/28F	1979	Ex Bristol, 1983
224	TWS914T	Bristol VRT/SL3/6LXB	Eastern Coach Works	H43/28F	1979	Ex Bristol, 1983

Cheltenham & Gloucester operate thirteen Leyland Olympians which are currently based at all operating units. Here we see 118, LWS36Y, one of four with the Stroud Valleys operation, heading for Forest Green on service 93. *Robert Edworthy*

225-231	Bristol VRT/SL3/680 (6LXB)	Eastern Coach Works	H43/31F	1981	Ex Bristol, 1983

225	DHW350W	**227**	EWS740W	**229**	EWS746W	**230**	EWS748W	**231**	EWS751W
226	DHW352W	**228**	EWS743W						

301-312	Leyland National 11351A/1R(DAF)		B52F*	1977-79	Ex Bristol, 1983 *301 is B52DL(Variable)

301	467WYA	**303**	TAE641S	**306**	PHW988S	**308**	TAE642S	**311**	SAE756S
302	YFB973V	**305**	PHW989S	**307**	SAE752S	**310**	VEU231T	**312**	TAE644S

314	YFB972V	Leyland National 11351A/1R(DAF)		B52F	1979	Ex Bristol, 1983

361-375	Leyland National 2 NL116L11/1R		B52F*	1980	Ex Bristol, 1983 *368 is B52FL (variable)

361	AAE644V	**364**	AAE648V	**367**	AAE651V	**370**	AAE660V	**373**	BHY997V
362	HIL6075	**365**	AAE649V	**368**	YJV806	**371**	AAE665V	**374**	BHY998V
363	511OHU	**366**	AAE650V	**369**	AAE659V	**372**	BHY996V	**375**	BOU6V

376	ARN892Y	Leyland National 2 NL116HLXB/1R		DP52F	1983	Ex Ribble, 1994
377	RHG880X	Leyland National 2 NL116AL11/1R		B52F	1982	Ex Ribble, 1994
378	NHH382W	Leyland National 2 NL116AL11/1R		B52F	1981	Ex Ribble, 1994
379	SHH389X	Leyland National 2 NL116AL11/1R		B52F	1982	Ex Ribble, 1994
380	RRM385X	Leyland National 2 NL116AL11/1R		B52F	1981	Ex Ribble, 1995
381	SNS825W	Leyland National 2 NL116AL11/1R		B52F	1980	Ex Ribble, 1995
382	KHH376W	Leyland National 2 NL116AL11/1R		B52F	1980	Ex Ribble, 1995
383	WAO397Y	Leyland National 2 NL116HLXB/1R		B52F	1983	Ex Busways, 1996
391	LFR860X	Leyland National 2 NL106AL11/1R		B44F	1981	Ex Ribble, 1995
392	LFR861X	Leyland National 2 NL106AL11/1R		B44F	1981	Ex Midland Red, 1995
393	LFR873X	Leyland National 2 NL106AL11/1R		B44F	1981	Ex Midland Red, 1995

401-409	Volvo B10M-55	Alexander PS	DP48F	1995	

401	N401LDF	**403**	N403LDF	**405**	N405LDF	**407**	N407LDF	**409**	N409LDF
402	N402LDF	**404**	N404LDF	**406**	N406LDF	**408**	N408LDF		

500	VAE499T	Leyland National 10351B/1R		B44F	1978	Ex Bristol, 1983
501w	VAE501T	Leyland National 10351B/1R		B44F	1979	Ex Bristol, 1983
503	VAE507T	Leyland National 10351B/1R		B44F	1979	Ex Bristol, 1983
533	G533LWU	Volvo B10M-60	Plaxton Paramount 3500 III	C48FT	1990	Ex Wallace Arnold, 1993
534	G534LWU	Volvo B10M-60	Plaxton Paramount 3500 III	C48FT	1990	Ex Wallace Arnold, 1993
546	G546LWU	Volvo B10M-60	Plaxton Paramount 3500 III	C48FT	1990	Ex Wallace Arnold, 1993
547	G547LWU	Volvo B10M-60	Plaxton Paramount 3500 III	C48FT	1990	Ex Wallace Arnold, 1993
548	G548LWU	Volvo B10M-60	Plaxton Paramount 3500 III	C48FT	1990	Ex Wallace Arnold, 1993

631-645	Ford Transit 190	Alexander AM	B16F	1985	

631	C631SFH	**636**	C636SFH	**640**	C640SFH	**642**	C642SFH	**645**	C645SFH
633	C633SFH	**637**	C637SFH	**641**	C641SFH	**643**	C643SFH		

651	C651XDF	Mercedes-Benz L608D	Alexander AM	B20F	1986	
659	C659XDF	Mercedes-Benz L608D	Alexander AM	B20F	1986	
677	F677PDF	Mercedes-Benz 709D	PMT	B25F	1988	
678	F311DET	Mercedes-Benz 709D	Reeve Burgess Beaver	B25F	1988	Ex Reeve Burgess demonstrator, 1989

679-684	Mercedes-Benz 709D	PMT	B25F	1989	

679	G679AAD	**681**	G681AAD	**682**	G682AAD	**683**	G683AAD	**684**	G684AAD
680	G680AAD								

Opposite: **Stroud Valley's operation is based on the London Road garage in Stroud. Here are two of their units. Leyland National 377, RHG880X is one of several of the mark 2 Nationals to be moved south from Ribble in 1994. It is seen heading for Wotton-under-Edge. As part of the Stagecoach modernisation programme, old minibuses are being replaced with Mercedes-Benz 709's with Alexander Sprint bodies. Some four hundred are being allocated this year. One of the 1995 intake, 704, M704JDG is shown here.**

Stagecoach
40 Wotton-under-Edge
RHG 880X

34 MASON ROAD
via Summer Cres.
Stagecoach
Stagecoach
Tourist Information
M704 JDG

686-703

Mercedes-Benz 709D | Alexander Sprint | B25F | 1994

686	L686CDD	**690**	L690CDD	**694**	L694CDD	**697**	M697EDD	**701**	M701EDD
687	L687CDD	**691**	L691CDD	**695**	L695CDD	**698**	M698EDD	**702**	M702EDD
688	L688CDD	**692**	L692CDD	**696**	L696CDD	**699**	M699EDD	**703**	M703EDD
689	L689CDD	**693**	L693CDD						

704-717

Mercedes-Benz 709D | Alexander Sprint | B25F* | 1995 | *704 is DP25F

704	M704JDG	**707**	M707JDG	**710**	M710JDG	**713**	M713FMR	**716**	N716KAM
705	M705JDG	**708**	M708JDG	**711**	M711FMR	**714**	M714FMR	**717**	N717KAM
706	M706JDG	**709**	M709JDG	**712**	M712FMR	**715**	M715FMR		

718-735

Mercedes-Benz 709D | Alexander Sprint | B25F* | 1996 | *731-5 are DP25F

718	N718RDD	**722**	N722RDD	**726**	N726RDD	**730**	N730RDD	**733**	N733RDD
719	N719RDD	**723**	N723RDD	**727**	N727RDD	**731**	N731RDD	**734**	N734RDD
720	N720RDD	**724**	N724RDD	**728**	N728RDD	**732**	N732RDD	**735**	N735RDD
721	N721RDD	**725**	N725RDD	**729**	N729RDD				

801	K801OMW	Mercedes-Benz 811D	Wright NimBus	B33F	1993	
802	K802OMW	Mercedes-Benz 811D	Wright NimBus	B33F	1993	
803	L803XDG	Mercedes-Benz 811D	Marshall C16	B33F	1993	
804	L804XDG	Mercedes-Benz 811D	Marshall C16	B33F	1993	
805	L805XDG	Mercedes-Benz 811D	Marshall C16	B33F	1993	
806	L806XDG	Mercedes-Benz 811D	Marshall C16	B33F	1993	
807	L330CHB	Mercedes-Benz 811D	Marshall C16	B33F	1993	Ex Red & White, 1994
808	K308YKG	Mercedes-Benz 811D	Wright NimBus	B33F	1992	Ex Red & White, 1995

831-845

Volvo B6-9.9M | Alexander Dash | B40F | 1994

831	L831CDG	**834**	L834CDG	**837**	L837CDG	**840**	L840CDG	**843**	M843EMW
832	L832CDG	**835**	L835CDG	**838**	L838CDG	**841**	L841CDG	**844**	M844EMW
833	L833CDG	**836**	L836CDG	**839**	L839CDG	**842**	L842CDG	**845**	M845EMW

846	L248CCK	Volvo B6-9.9M	Alexander Dash	DP40F	1993	Ex Ribble, 1995
847	M847HDF	Volvo B6-9.9M	Alexander Dash	B40F	1994	
848	L709FWO	Volvo B6-9.9M	Alexander Dash	B40F	1994	Ex Red & White, 1995
849	L710FWO	Volvo B6-9.9M	Alexander Dash	B40F	1994	Ex Red & White, 1995
850	L711FWO	Volvo B6-9.9M	Alexander Dash	B40F	1994	Ex Red & White, 1995
851	L712FWO	Volvo B6-9.9M	Alexander Dash	B40F	1994	Ex Red & White, 1995

Circle-Line:

1065	KKW65P	Leyland Leopard PSU3C/4R	Alexander AY	DP43DL	1975	Ex South Yorkshire's Transport, 1992
1102	E102OUH	Freight Rover Sherpa	Carlyle	B20F	1987	Ex Red & White, 1994
1114	NFB114R	Bristol VRT/SL3/6LXB	Eastern Coach Works	H43/27D	1976	Ex City Line, 1993
1137	AKV137V	Leyland Atlantean AN68A/2R	Alexander AL	H49/37F	1980	Ex Busways, 1996
1247	SCN247S	Leyland Atlantean AN68A/2R	Alexander AL	H49/37F	1978	Ex Busways, 1996
1250	SCN250S	Leyland Atlantean AN68A/2R	Alexander AL	H49/37F	1978	Ex Busways, 1996
1255	SCN255S	Leyland Atlantean AN68A/2R	Alexander AL	H49/37F	1978	Ex Busways, 1996
1256	SCN256S	Leyland Atlantean AN68A/2R	Alexander AL	H49/37F	1978	Ex Busways, 1996
1264	SCN264S	Leyland Atlantean AN68A/2R	Alexander AL	H49/37F	1978	Ex Busways, 1996
1310	REU310S	Bristol VRT/SL3/6LXB	Eastern Coach Works	H43/28F	1977	
1468	PPH468R	Bristol VRT/SL3/501	Eastern Coach Works	H43/31F	1977	Ex Swanbrook, Cheltenham, 1991
1469	A469TUV	Leyland Cub CU335	Wadham Stringer Vanguard	B21FL	1984	Ex LB of Wandsworth, 1992
1480	C480BFB	Ford Transit 190	Dormobile	B16F	1986	Ex Badgerline, 1993
1499	C499BFB	Ford Transit 190	Dormobile	B16F	1986	Ex Badgerline, 1993
1505	JOX505P	Leyland National 11351A/1R		B49F	1976	Ex Stagecoach Midland Red, 1995
1511	LUL511X	Leyland Cub CU335	Wadham Stringer Vanguard	B21FL	1982	Ex London Residuary Body, 1994
1515	PEU515R	Bristol VRT/SL3/6LXB	Eastern Coach Works	H43/31F	1977	Bristol, 1983
1545	DNE545Y	Dodge G10	Wadham Stringer	DP30CL	1983	Ex Community Routes, Hattersley, 1993
1581	D581VBV	Freight Rover Sherpa	Dormobile	B16F	1988	Ex Lane, Churchdown, 1994
1603	NFB603R	Leyland National 11351A/1R		B52F	1977	Ex Bristol, 1983
1604	D604HTC	Iveco 79-14	Robin Hood	B30FL	1987	Ex Gloucestershire CC, 1992
1605	D605HTC	Iveco 79-14	Robin Hood	B30FL	1987	Ex Gloucestershire CC, 1992
1606	D606HTC	Iveco 79-14	Robin Hood	B30FL	1987	Ex Gloucestershire CC, 1992
1617	C617SFH	Ford Transit 190	Alexander AM	B16F	1985	
1621	C621SFH	Ford Transit 190	Alexander AM	B16F	1985	
1626	C626SFH	Ford Transit 190	Alexander AM	B16F	1985	
1639	C639SFH	Ford Transit 190	Alexander AM	B16F	1985	

Twenty-one Volvo B6 buses are operated with four having joined the fleet from Red and White and one with high-back seating from Ribble. One of the former Red & White vehicles, 850, L711FWO is seen on Gloucester Citybus service 10. All but five of the B6s are based at Gloucester.
Robert Edworthy

1663	E663JAD	MCW MetroRider MF150/43	MCW	B25F	1987	
1665	E665JAD	MCW MetroRider MF150	MCW	B25F	1987	
1667	E667JAD	MCW MetroRider MF150/43	MCW	B25F	1987	
1676	E676KDG	MCW MetroRider MF150/61	MCW	DP25F	1988	
1693	C693VAD	Leyland Cub CU435	Wadham Stringer Vanguard	B32FL	1986	Ex Gloucestershire CC, 1992
1694	C694VAD	Leyland Cub CU435	Wadham Stringer Vanguard	B32FL	1986	Ex Gloucestershire CC, 1992
1696	C696VAD	Leyland Cub CU435	Wadham Stringer Vanguard	B32FL	1986	Ex Gloucestershire CC, 1992
1697	C697VAD	Leyland Cub CU435	Wadham Stringer Vanguard	B32FL	1986	Ex Gloucestershire CC, 1992
1710	SND710X	Leyland Tiger TRCTL11/3R	Plaxton Supreme V Express	C53F	1982	Ex Cumberland, 1994
1713	WLT713	Leyland Tiger TRCTL11/3RH	Duple Laser 2	C46F	1984	Ex Cumberland, 1994
1726	LHT726P	Bristol VRT/SL3/6LXB	Eastern Coach Works	H43/27D	1976	Ex City Line, 1993
1738	C738CUC	Leyland Cub CU335	Wadham Stringer	B21FL	1986	Ex LB of Wandsworth, 1992
1740	MOU740R	Bristol VRT/SL3/501(6LXB)	Eastern Coach Works	H43/27D	1975	Ex City Line, 1993
1754	SAE754S	Leyland National 11351A/1R(DAF)		B52F	1978	Ex Bristol, 1983
1804	6804VC	Leyland Tiger TRCTL11/3RH	Plaxton Paramount 3200 II	C51F	1986	Ex Stagcoach Midland Red, 1996
1838	LOA838X	Leyland Leopard PSU3F/4R	Willowbrook 003	C49F	1982	Ex Vanguard, Bedworth, 1995
1977	DSD977V	Seddon Pennine 7	Alexander AT	C49F	1979	Ex Western Scottish, 1995

Operating Units:

Circle-Line:	All 1xxx numbers
Cirencester:	117, 204/15, 401-3, 679/83
Gloucester Citybus:	114/21-3, 213/23/4, 301/3/5/7/8/10/1/4, 391/3, 503, 680-2/4/7/8/99, 701-3/23-7/33/4, 831-42/8-51.
Stroud Valleys:	115/8-20, 226-31, 362/3/8/76-9/81/2, 500, 645/51/9/77, 704/18-22/35, 803-8
Swindon & District:	101/2/4-6/8-13/6/24, 213/8/9/22, 369-73/80/3, 501, 711-7, 801/2/43-5.
Cheltenham District:	Remainder

Previous Registrations:

467WYA	TAE645S	6804VC	WVT618, C473CAP	YJV806	AAE658V
511OUH	AAE647V	HIL6075	AAE646V	WLT713	B108HAO

CHENEY COACHES

Cheney Coaches Ltd, Cheney House, Thorpe Drive, Thorpe Ind Est, Banbury, Oxfordshire, OX16 8UZ

URH13R	Bedford YMT	Plaxton Supreme III Express	C53F	1977	Ex Dorset CC, 1994
9785SM	Bedford YMT	Plaxton Supreme IV	C53F	1979	Ex Tappins, Didcot, 1994
PDN873	Bedford YMT (Cummins)	Plaxton Supreme IV	C53F	1979	Ex Tappins, Didcot, 1994
VPF742	Bedford YMT (Cummins)	Plaxton Supreme IV	C53F	1979	Ex Tappins, Didcot, 1994
OGL849	Bedford YMT	Plaxton Supreme IV	C53F	1979	Ex Tappins, Didcot, 1993
JDX574V	Bedford YMT	Duple Dominant II Express	C53F	1979	Ex Sanders, Holt, 1995
RFC443W	Bedford YLQ	Plaxton Supreme IV Express	C45F	1980	Ex Dorset Police, 1994
B586EGT	Renault-Dodge S56	Dormobile	B22FL	1985	Ex LB Merton, 1993
C354FBO	Bedford YNV Venturer	Plaxton Paramount 3200 II	C57F	1985	Ex Billies, Mexborough, 1995
YSU975	TAZ D3200	TAZ Dubrava	C49FT	1985	Ex Safeguard, Guildford,1993
USU800	Iveco 315	Caetano Algarve	C28F	1987	Ex Scancoaches, North Acton, 1993
URT682	Bedford YNV Venturer	Plaxton Paramount 3200 II	C55F	1987	Ex Roy Brown, Builth Wells, 1994
E701GNH	Iveco 315,	Caetano Algarve	C28F	1987	Ex Silver Fern, Gatwick, 1995
E175TWW	Bedford YNV Venturer	Plaxton Paramount 3200 III	C57F	1988	Ex Blue Iris, Nailbea, 1994
G22UWL	Ford Transit VE6	Jubilee	M12	1990	
G711VRY	TAZ D3500	TAZ Dubrava	C49FT	1989	Ex SMC, Garston, 1996
H787JFC	Ford Transit VE6	Ford	M8	1991	
H562FLE	Mercedes-Benz 609D	North West Coach Sales	C19FL	1990	Ex Capital, Heathrow, 1996
K373HHK	Ford Transit	Ford	M14	1993	

Previous Registrations:

9785SM	YAN817T	PDN873	YAN818T	USU800	D835CNV
NSU914	KPR329W	RFC443W	KPR329W, NSU914	VPF742	YAN819T
OGL849	YAN823T	URT682	D626BVJ	YSU975	F794TBC, DSK558, F199UPC

Livery: White, blue and red

Representing Cheney Coaches is 9785SM, a Bedford YMT with Plaxton Supreme IV body and one of four acquired from nearby Tappins in 1994. The picture was taken in Banbury as the vehicle was preparing to start it's afternoon school duty. *Tony Wilson*

CHILTERN QUEENS

Chiltern Queens Ltd, Long Toll, Greenmore Hill, Woodcote, Oxfordshire, RG8 0RP

OJO835M	Leyland Leopard PSU3B/4R	Plaxton Derwent	B55F	1974	
VBW581	Leyland Leopard PSU5A/4R	Plaxton Supreme III	C57F	1976	
RFC10T	Leyland Leopard PSU3E/4R	Duple Dominant II Express	C49F	1978	Ex Oxford Bus Company, 1989
RFC12T	Leyland Leopard PSU3E/4R	Duple Dominant II Express	C49F	1978	Ex Oxford Bus Company, 1990
WUD815T	Leyland Leopard PSU3E/4R	Duple Dominant II Express	C49F	1978	Ex Oxford Bus Company, 1990
591STT	Leyland Leopard PSU3E/4R	Plaxton Supreme IV Express	C53F	1979	
YFC18V	Leyland Leopard PSU3E/4R	Duple Dominant II Express	C49F	1979	Ex Oxford Bus Company, 1991
BBW20V	Leyland Leopard PSU3E/4R	Duple Dominant II Express	C49F	1979	Ex Oxford Bus Company, 1993

Chiltern Queens continues to gather former Oxford Bus Company coaches into their fleet including the first two Leyland Tigers to arrive here. Pictured outside Reading rail station about to work the service to Woodcote is F344TSC, a Mercedes-Benz 811D with Alexander Sprint bodywork. This carries a livery of two-tone green and white. In contrast is B911SPR, pictured in Llandudno. This Volvo B10M carries a two-tone blue livery on it's Plaxton Paramount body.
Ralph Stevens

Four Mercedes-Benz make up the Chiltern Queens minibus fleet. Running the main services along with F344TSC on the previous page is F986TTF, an Optare-bodied Mercedes-Benz 811. The normal Star-Rider body replaced the standard Mercedes-Benz cowl usually retained by the bodybuilder. Again, Reading rail station provides the location. *Philip Lamb*

BBW22V	Leyland Leopard PSU3E/4R	Duple Dominant II Express	C49F	1979	Ex Oxford Bus Company, 1992
2969HJ	Leyland Leopard PSU3E/4R	Plaxton Supreme IV	C53F	1979	Ex Regis Coaches, Challow, 1996
MUD25W	Leyland Leopard PSU3F/4R	Duple Dominant II Express	C49F	1981	Ex Oxford Bus Company, 1993
PPJ65W	Leyland Leopard PSU5C/4R	Wadham Stringer Vanguard	B54F	1982	Ex M o D, 1993
PJH582X	Leyland Leopard PSU3E/4R	Plaxton Supreme III	C53F	1982	
EBW106Y	Leyland Tiger TRCTL11/3R	Duple Dominant IV Exp	C51F	1983	Ex Oxford Bus Company, 1995
EBW107Y	Leyland Tiger TRCTL11/3R	Duple Dominant IV Exp	C51F	1983	Ex Oxford Bus Company, 1996
B911SPR	Volvo B10M-61	Plaxton Paramount 3200 II	C53F	1985	Ex Excelsior, Bournemouth, 1987
TSV804	Volvo B10M-61	Jonckheere Jubilee	C49FT	1986	Ex Gunton, Ongar, 1994
C644SJM	Volvo B10M-61	Plaxton Paramount 3200 II	C53F	1986	
C114PUJ	Volvo B10M-61	Caetano Algarve	C49FT	1986	Ex Hughes, Llanfair Caereinion, 1993
D504NWG	Mercedes-Benz L608D	Alexander AM	B20F	1986	Ex RoadCar, 1995
D506NWG	Mercedes-Benz L608D	Alexander AM	B20F	1986	Ex RoadCar, 1995
D262HFX	Volvo B10M-61	Plaxton Paramount 3200 III	C53F	1987	Ex Excelsior, Bournemouth, 1988
D34ENH	Volvo B10M-61	Duple 340	C55F	1987	Ex Country Lion, Northampton, 1994
E533PRU	Volvo B10M-61	Plaxton Paramount 3200 III	C48FT	1987	
F344TSC	Mercedes-Benz 811D	Alexander AM	DP29F	1988	Ex Challenger, Bridgnorth, 1992
F986TTF	Mercedes-Benz 811D	Optare StarRider	B33F	1989	Ex Davron Travel, Caversham, 1991
H788RWJ	Scania K93CRB	Plaxton Paramount 3200 III	C55F	1990	

Previous Registrations:

2969HJ	EBM443T	PPJ65W	50AC08	TSV804	C28GNK
591STT	UUD623T	C114PUJ	C690KDS, SEL4X	VBW581	SFC32P

Livery: Red and white; two-tone green and white (coaches)

N N CRESSWELL

N N Cresswell, M S Shephard and E C Everatt, Coach House, Worcester Road, Evesham, Worcestershire, WR11 4RA

HVN601N	Bedford YRQ	Plaxton Elite III	C45F	1975	Ex Owen's Motors, Knighton, 1984
PJI8917	Bedford YMT	Plaxton Supreme III	C53F	1978	Ex Westours, Pershore, 1993
XCJ750T	Bedford YMT	Plaxton Supreme IV Express	C53F	1979	
KPT583T	Bedford YMT	Plaxton Supreme IV	C53F	1979	Ex Norrie, New Deer, 1988
AUJ715T	Bedford YMT	Duple Dominant II	C53F	1979	Ex Whittle, Highley, 1982
BWE196T	Bedford YMT	Duple Dominant II	C53F	1979	Ex Smith, Evesham, 1995
FUJ938V	Bedford YMT	Duple Dominant II	C53F	1980	Ex Owen, Oswestry, 1988
FUJ950V	Bedford YMT	Plaxton Supreme IV	C53F	1980	Ex Whittle, Highley, 1983
JRB663V	Bedford YMT	Plaxton Supreme IV	C53F	1980	Ex Kime, Folkingham, 1993
BGR630W	Bedford YLQ	Duple Dominant II Express	C45F	1980	Ex Bond Bros, Willington, 1994
LIB1797	Bedford YNT	Plaxton Supreme IV	C53F	1981	Ex Hilo, Sandy, 1991
LCJ626Y	Bedford YNT	Plaxton Paramount 3200 E	C53F	1983	
B180TVJ	Bedford YNT	Plaxton Paramount 3200 II	C53F	1985	
C175HYD	Bedford YNV Venturer	Duple Laser 2	C57F	1986	Ex Coleman, Yeovil, 1988
C48HKK	Bedford YNV Venturer	Plaxton Paramount 3200 II	C53F	1986	Ex Camden, Sevenoaks, 1993
D827PUK	Freight Rover Sherpa	Carlyle	B18F	1987	Ex Ribble, 1992
E403WAM	Renault-Dodge S56	Reeve Burgess Beaver	DP25F	1987	Ex Somerbus, Paulton, 1996
E388FLD	Bedford YNV Venturer	Plaxton Paramount 3200 III	C53F	1988	Ex Capital, West Drayton, 1995
F387KVJ	Freight Rover Sherpa	Freight Rover	M16	1988	
J2NNC	MAN 10-180	Caetano Algarve II	C35F	1991	
L2NNC	Mercedes-Benz 709D	Plaxton Beaver	B25F	1994	

Previous Registrations:
LIB1797 KKW451W
PJI8917 URX524S, RJB895

Livery: White and blue

Worcester service 552 to Longlartin, location of a major maximum security prison, is operated by N N Cresswell and pictured on a wet day in Evesham is the newest minibus in the fleet, L2NNC. The 1994 Plaxton Beaver was joined in 1996 by a similarly designed Reeve Burgess model based on a Dodge chassis and originated with Thamesdown. *Richard Godfrey*

COTTRELL'S

BN & ER Cottrell, St Michaels Close, Mill End, Mitcheldean, GL17 0HP

u	EAD122T	Leyland Fleetline FE33ALR	Northern Counties	H47/36F	1979	
	GDF332V	Volvo B58-56	Plaxton Supreme IV	C51F	1979	
	GBU2V	MCW Metrobus DR101/6	MCW	H43/30F	1979	Ex GM Buses, 1986
	GBU6V	MCW Metrobus DR101/6	MCW	H43/30F	1979	Ex GM Buses, 1986
	PAD806W	Leyland Leopard PSU3E/4R	Plaxton Supreme IV Express	C51F	1981	
	NDE147Y	Leyland Tiger TRCTL11/2R	Plaxton Paramount 3200E	C53F	1983	Ex Horlock, Northfleet, 1988
	C474CAP	Leyland Tiger TRCTL11/3RH	Plaxton Paramount 3200 II	C51F	1986	Ex Thames Transit, 1991
	D803NBO	Leyland Tiger TRCTL11/3RH	Plaxton Paramount 3500 II	C51FT	1987	Ex Crown Coaches, Bristol, 1991
	D160UGA	Leyland Lion LDTL11/1R	Alexander RH	DPH49/23F	1987	Ex Clydeside 2000, 1994
	E688UNE	Leyland Tiger TRCTL11/3RZ	Plaxton Paramount 3200 III	C53F	1988	Ex Shearings, 1992
	E478AFJ	Leyland Tiger TRCTL11/3RZ	Plaxton Paramount 3500 III	C53F	1988	Ex Loverings, Combe Martin, 1995
	F309RMH	Leyland Tiger TRBTL11/2RP	Duple 300	B55F	1989	Ex Rover Bus Service, Chesham, 1993
	F66SMC	Leyland Cub LBM6T/2RA	Wadham Stringer Vanguard	B39F	1989	Ex Arrow Coaches, Bristol, 1993
	F183UFH	Leyland Tiger TRCTL11/3ARZ	Plaxton Paramount 3200 III	C57F	1989	

Previous Registrations:

D160UGA	D852RDS, 705DYE
C474CAP	YDG616

Livery: Maroon and cream

Opposite: **Cotterell's started as carriers in the late 1880s, running twice weekly to Gloucester which remains the target for its now seven-days-a-week service. Until the 1920s horse-drawn vehicles were used. Somewhat advanced from those times are** *(top)* **F309RMH, a Duple-bodied Leyland Tiger originally with Rover Bus and** *(bottom)* **former Clydeside Leyland Lion D160UGA, with Alexander bodywork. The weekday service to Gloucester is hourly and crewe operation is still employed Monday to Friday from Ruardean which alternates with Cinderford as the starting point.**

Below: **Seen leaving the bus station is F183UFH, a 12-metre Leyland Tiger with Plaxton Paramount 3200 bodywork and Cotterell's most recent new purchase.** *Les Peters*

C24
CINDERFORD VIA
HUNTLEY & MITCHELDEAN
COTTRELL'S
COTTRELL'S
Leyland
F309 RMH

COTTRELL'S COACHES
RUARDEAN
VIA
HUNTLEY & MITCHELDEAN
25
D160 UGA

DE LUXE

De Luxe Coach Services Ltd, 2 Mancetter Road, Atherstone, Warwickshire, CV9

RPT293K	Leyland Leopard PSU3B/4R	Alexander AY	DP47F	1972	Ex Reed, Sunnyside, 1992
YSD340L	Leyland Leopard PSU3E/4R	Alexander AY	DP53F	1973	Ex Lockley, Stafford, 1994
DRB61T	Leyland Leopard PSU5C/4R	Plaxton Supreme IV	C57F	1979	Ex Dons, Dunmow, 1992
EWW208T	Leyland Leopard PSU3E/4R	Plaxton Supreme IV	C53F	1979	Ex Spencer, New Ollerton, 1984
LDS381V	Leyland Leopard PSU3E/4R	Plaxton Supreme IV	C53F	1980	Ex Garelochhead CS, 1981
LBO10X	Volvo B10M-61	Plaxton Viewmaster IV	C53F	1981	Ex Bebb, Llantwit Fardre, 1985
YEH182X	Volvo B10M-61	Duple Dominant IV	C53F	1981	Ex Baker, Biddulph, 1985
XVC230X	Leyland Leopard PSU5C/4R	Plaxton Supreme IV	C57F	1982	
NXI918	Volvo B10M-60	Plaxton Paramount 3500 III	C49FT	1989	Ex Allison's Coaches, Dunfermline, 1994

Previous Registrations:

YEH182X	FHS738X, 3471RU	NXI918	F998HGE, USU907, F510NSH

Livery: Red, white and black

Pictured at Coventry are two examples of Plaxton bodywork showing differing styles. On the left is XVC230X, a Supreme IV which is based on a Leyland Leopard, while on the right is LBO11X, a viewmaster based on a Volvo B10M. During 1996 this vehicle was sold, though an identical vehicle, LBO10X, is still used. The viewmaster was built from 1977-82 and was, in effect, a high-floor Supreme. The increasing influx of high-floor continental bodywork gave rise to this solution which was more successful on the later Tiger/B10M/DAF chassis than the Leopard/B58 available at the outset. Over 200 were built. *David Cole*

DRM

D R Morris, Coach Garage, Broadbridge, Bromyard, Herefordshire, HR7 4NT

Depot: Stourport Road, Bromyard.

MOI3512	Volvo B58-61	Plaxton Elite III Express	C53F	1972	Ex Down, Mary Tavy, 1986
MOI5633	AEC Reliance 6U3ZR	Plaxton Elite III Express	C49F	1972	
PGU995R	Volvo B58-56	Plaxton Supreme III	C49F	1976	Ex West Kingsdown Coaches, 1996
MOI1793	Leyland Leopard PSU3E/4R	Plaxton Supreme III Express	C53F	1977	
MOI5055	Leyland National 11351A/1R(Volvo)		B49F	1978	Ex Alpine, Llandudno, 1993
LBU781V	Leyland Leopard PSU3E/4R	Plaxton Supreme IV Express	C49F	1979	Ex A-Line, Bedworth, 1996
MOI9565	Volvo B10M-61	Jonckheere Jubilee P50	C53FT	1983	Ex Budden's, Woodfalls, 1988
MOI4000	Volvo B10M-61	East Lancashire (1995)	DP53F	1990	Ex Classic, Paignton, 1995

Previous Registrations:

LBU781V	SPT219V, GSU489	MOI3565		MOI5633	OKX49L
MOI1793	OVJ400R	MOI4000	G68RGG	MOI7000	
MOI3512	KEG773L, 5635CD	MOI5055	RKA871T	MOI9565	NNV554Y

Livery: White

DRM of Bromyard operates MOI4000, a re-bodied Volvo B10M coach chassis. Now fitted with an East Lancashire body, and fitted with high-back seating, the vehicle is used on Hereford services. It is seen in the County town waiting time before its return journey to Ledbury. David Morris' father, Bill Morris, who featured in the previous edition, retired in the summer of 1996. *Richard Eversden*

DUDLEY'S COACHES

Dudley Coaches Ltd, Poplar Garage, Radford, Inkberrow, Worcestershire, WR7 4LS

UWP96R	Ford R1114	Plaxton Supreme III Express	C53F	1976	
TWP97V	Ford R1114	Plaxton Supreme IV Express	C53F	1980	
JWB847W	Leyland Leopard PSU5D/5R	Plaxton Supreme IV	C57F	1980	Ex Royal, Redditch, 1986
MIW1607	Volvo B58-61	Van Hool Alizée	C52F	1980	Ex Warner, Tewkesbury, 1987
UOI7274	Volvo B58-56	Plaxton Supreme IV Express	C48FT	1980	Ex DRM, Bromyard, 1988
KHL460W	Volvo B58-56	Plaxton Supreme IV	C53F	1981	Ex Woodstones, Kidderminster, 1988
CIL9223	Leyland Leopard PSU5C/4R	Plaxton Supreme IV	C57F	1981	Ex Royal, Redditch, 1986
STO244X	Leyland Leopard PSU3F/5R	Plaxton Supreme IV	C53F	1981	Ex Royal, Redditch, 1986
KUY98X	Ford R1114	Plaxton Supreme V Express	C53F	1982	
A722BAB	Volvo B10M-61	Duple Caribbean	C51FT	1984	
B951TKV	Leyland Tiger TRCTL11/3RZ	Plaxton Paramount 3500 II	C53F	1985	Ex Richardson, Midhurst, 1994
C685MWJ	Volvo B10M-61	Plaxton Paramount 3200 II	C57F	1986	Ex Bere Regis & District, 1994
C319UFP	Volvo B10M-61	Plaxton Paramount 3200 II	C57F	1986	Ex Crawford, Neilston, 1989
D225LWY	Volvo B10M-61	Plaxton Paramount 3500 III	C53F	1987	Ex Wallace Arnold, 1993
E675UNE	Leyland Tiger TRCTL11/3RZ	Plaxton Paramount 3200 II	C53F	1988	Ex Essex Bus (Thamesway), 1996
F952ENH	Dennis Javelin 12SDA1907	Plaxton Paramount 3200 III	C53F	1988	Ex Wainfleet, Nuneaton, 1996
F940WFA	Volvo B10M-61	Plaxton Paramount 3200 III	C52F	1989	Ex Baker, Biddulph, 1996

Previous Registrations:

B951TKV	684DYX
C685MWJ	OJT568
CIL9223	BGS288X
F940WFA	F487LHO, 3353RU
F952ENH	F376MUT, MIW5793
MIW1607	BSF337W, 9246WF, SFH655W
UOI7274	ECJ700W

Livery: Cream and two-tone green

Representing Dudley's Coaches is Leyland Leopard CIL9223, one of three in the fleet with Plaxton Supreme IV bodies, all arriving with the purchase of the Royal Motorways business in 1986. It was seen on layover after working the 350 service to Redditch. 1996 has been a busy year with several additions to the fleet, including the first Dennis Javelin. Only two vehicle in this fleet carry bodywork other than Plaxton. *Robert Edworthy*

DUKES TRAVEL

KG & PK Bevan, Ferndale, Edge End, Coleford, GL16 7ED

Depot : Laker's Road, Berry Hill

509RHU	Volvo B58-61	Plaxton Supreme IV	C57F	1980	Ex Liddell, Bristol, 1994
3012WF	Leyland Leopard PSU5D/5R	Wadham Stringer Vanguard	B50F	1981	Ex MoD, 1993
TAD24W	Leyland Tiger TRCTL11/3R	Plaxton Supreme IV	C50F	1981	Ex Warner, Tewkesbury, 1991
MYO486X	Leyland Leopard PSU5G/4R	Wadham Stringer Vanguard	B52F	1982	Ex MoD, 1991
FSU803	Volvo B10M-61	Van Hool Alizée	C53F	1984	Ex Barratt, Nantwich, 1995
116XYD	Leyland Tiger TRCTL11/3R	Plaxton Paramount 3500	C55F	1984	Ex Lewis, Greenwich, 1988
666VHU	Leyland Royal Tiger RT	Van Hool Alizée	C53F	1984	Ex Tellings Golden Miller, Byfleet, 1992
B460WTC	Ford Transit 190	Dormobile	B16F	1985	Ex Fosseway, Grittleton, 1992
216TYC	Volvo B10M-61	Jonckheere Jubilee	C51FT	1987	Ex John Morrow, Glasgow, 1995
E132KYW	MCW MetroRider MF150/38	MCW	B25F	1987	Ex Rhondda, 1995
E137KYW	MCW MetroRider MF150/38	MCW	B25F	1987	Ex Rhondda, 1995

Previous Registrations:

116XYD	A614HGY	509RHU	MDS230V	MYO486X	O4RN34
216TYC	E218GNV	666VHU	B551TWR	TAD24W	SDG25W, 4529WF
3012WF	50AC05	FSU803	A643UGD		

Livery: White, green and orange

Dukes Travel have been established some 22 years. The current father and son partnership operates local services in the western Forest of Dean as well as a Saturday service to Gloucester. Previously with Rhondda and now with Dukes Travel are a pair of MCW MetroRiders that were delivered new to London Buses in 1987. Pictured heading for Ross on Wye is E132KYW. *Robert Edworthy*

GEOFF WILLETTS

FR Willetts & Co (Yorkley) Ltd, Main Road, Pillowell, Lydney, GL15 4QY

2464FH	Leyland National 11351/1R		B52F	1974	Ex East Yorkshire, 1994
BUH219V	Leyland National 11351A/1R (Vo)	East Lancs Greenway (1994)	B52F	1979	Ex Rhondda, 1993
890CVJ	Kässbohrer Setra S215HD	Kässbohrer	C49FT	1983	Ex Harris, Catshill, 1987
A291JDD	Leyland Tiger TRCTL11/3R	Plaxton Paramount 3200	C57F	1984	
E322UUB	Volvo B10M-61	Plaxton Paramount 3200 III	C53F	1988	Ex Wallace Arnold, 1991
G290XFH	Leyland Tiger TRCL10/3ARZA	Plaxton Paramount 3200 III	C57F	1989	
H937DRJ	Volvo B10M-60	Plaxton Paramount 3200 III	C53F	1991	Ex Shearings, 1995
L353MKU	Mercedes-Benz 814D	Plaxton Beaver	B33F	1993	

Previous Registrations:

2464FH GCY748N

890CVJ CRS327Y, 28JGH, UAB802Y

Livery: Mauve, maroon and red

The Willetts family business extends back to the 1930s and is now one of the most highly respected in Gloucestershire. Local services operate from Lydney to Monmouth through Coleford employing what are, possibly, the smartest Leyland Nationals in the country. Painted in the striking mauve, red and maroon scheme is 890CVJ, a Kässbohrer Setra S215HD. *Robert Edworthy*

GO WHITTLE

R A & D L Whittle; Corvedale Motor Co Ltd; M & M Coaches Ltd,
105 Coventry St, Kidderminster, Worcestershire, DY10 2BH

Depots : Central Garage, Alveley; Lion Lane, Cleobury Mortimer; High Street, Highley; Mouse Lane, Kidderminster and Fishmore Road, Ludlow.

1	ODN601	Leyland Leopard PSU5D/4R	Duple Dominant III	C57F	1980	
2	G102JNP	Dennis Javelin 12SDA1931	Plaxton Paramount 3200 III	C51F	1990	
3	F903RWP	Dennis Javelin 12SDA1917	Plaxton Paramount 3200 III	C57F	1989	
4	F904RWP	Dennis Javelin 12SDA1917	Plaxton Paramount 3200 III	C57F	1989	
5	JPY505	Bedford YMT	Plaxton Supreme III Express	C53F	1978	Ex Narburgh, Alveley, 1982
6	K6GOW	Dennis Dart 9.8SDL3017	Northern Counties Paladin	DP42F	1992	
7	F907RWP	Dennis Javelin 12SDA1917	Plaxton Paramount 3200 III	C57F	1989	
8	G108JNP	Dennis Javelin 12SDA1931	Plaxton Paramount 3200 III	C51F	1990	
9	M409PUY	Dennis Dart 9.8SDL3054	Northern Counties Paladin	DP42F	1995	
10	RCE510	Freight Rover Sherpa	Carlyle	B18F	1987	Ex Wigley, Herne Bay, 1991
11	G111JNP	Dennis Javelin 12SDA1931	Plaxton Paramount 3200 III	C51F	1990	
12	SIJ4712	Dennis Javelin 12SDA1917	Plaxton Paramount 3200 III	C51F	1989	
14	RPP514	Bedford YMT	Plaxton Supreme IV	C53F	1980	Ex Narburgh, Alveley, 1990
15	F115TWP	Mercedes-Benz 811D	Reeve Burgess Beaver	C33F	1989	
16	F116TWP	Freight Rover Sherpa	Crystals	M16	1989	

Go Whittle operate several services on behalf of Shropshire County Council as well as some commercial routes. Photographed in Kidderminster on a Bridgnorth journey is 55, XKH455, a recently re-registered Bedford YMT with Duple Dominant II bodywork. *Richard Godfrey*

17	H17GOW	Dennis Javelin 12SDA1929	Plaxton Paramount 3200 III	C53F	1991	
18	FUJ918V	Bedford YMT	Duple Dominant II	C57F	1980	
19	H19GOW	Dennis Javelin 12SDA1929	Plaxton Paramount 3200 III	C53F	1991	
20	L120CUY	Mercedes-Benz 709D	Plaxton Beaver	B23F	1994	
21	M421PUY	Dennis Javelin 12SDA2161	Plaxton Premiere 350	C57F	1995	
22	M422PUY	Dennis Javelin 12SDA2161	Plaxton Premiere 350	C57F	1995	
23	M423PUY	Dennis Javelin 12SDA2161	Plaxton Premiere 350	C49FT	1995	
24	FUJ924V	Bedford YMT	Duple Dominant II	C57F	1980	
25	M425PUY	Dennis Javelin 12SDA2161	Plaxton Premiere 350	C49FT	1995	
26	N26FUY	LDV 400	LDV	M16	1996	
27	N27FUY	LDV 400	LDV	M16	1996	
29	G29HDW	Dennis Javelin 12SDA1907	Duple 320	C57F	1990	Ex Bebb, Llantwit Fardre, 1992
30	G130JNP	Dennis Javelin 12SDA1930	Plaxton Paramount 3200 III	C53F	1990	
31	G131JNP	Dennis Javelin 12SDA1931	Plaxton Paramount 3200 III	C51F	1990	
32	M432PUY	Dennis Javelin 12SDA2161	Plaxton Premiere 350	C57F	1995	
33	K33GOW	Dennis Dart 9.8SDL3017	Northern Counties Paladin	DP42F	1992	
34	M434PUY	Dennis Javelin 12SDA2161	Plaxton Premiere 350	C49FT	1995	
36	M436PUY	Dennis Javelin 12SDA2161	Plaxton Premiere 350	C49FT	1995	
37	M437PUY	Dennis Javelin 12SDA2161	Plaxton Premiere 350	C49FT	1995	
39	F139TWP	Dennis Javelin 12SDA1917	Plaxton Paramount 3200 III	C53F	1989	
40	FUJ940V	Bedford YMT	Duple Dominant II	C57F	1980	Ex M & M, Kidderminster, 1981
41	URH341	Bedford YLQ	Plaxton Supreme III	C45F	1979	Ex Narburgh, Alveley, 1990
42	WTL642	Freight Rover Sherpa	Carlyle	B18F	1987	Ex Wigley, Herne Bay, 1991
44	F244SAB	Dennis Javelin 12SDA1917	Plaxton Paramount 3200 III	C57F	1989	
46	YFU846	Bedford YLQ	Duple Dominant II	C45F	1978	Ex M & M, Kidderminster, 1981
48	LSV548	Bedford YMT	Duple Dominant II	C53F	1979	Ex Narburgh, Alveley, 1990
51	F151TWP	Mercedes-Benz 811D	Reeve Burgess Beaver	C33F	1989	
52	F152TWP	Freight Rover Sherpa	Crystals	M16	1989	
52A	WVU152S	Bedford YMT	Duple Dominant II	C53F	1978	Ex Narburgh, Alveley, 1990
54	GBB254	Bedford YLQ	Duple Dominant II	C45F	1978	Ex M & M, Kidderminster, 1981
55	XKH455	Bedford YMT	Duple Dominant II	C57F	1980	Ex M & M, Kidderminster, 1981
87	TPJ287S	Bedford YMT	Duple Dominant II	C53F	1978	Ex Narburgh, Alveley, 1990

Previous Registrations:

GBB254	VNT23S	RCE510	D250OOJ	WTL642	D255OOJ
JPY505	ARB531T	RPP514	JPC783V, 6253VC, MWK826V	XKH455	FUJ948V
LSV548	YHB647T	SIJ4712	F112TWP	YFU846	VNT43S
ODN601	KUX221W	URH341	LJJ35P		

Livery: White, blue, green and yellow

Go-Whittle has become a committed user of the Dennis Javelin with Plaxton-bodied coaches being supplied new. Several Duple-bodied examples joined the fleet from Bebb in 1992 but most have been displaced by new Première 320s such as 25, M425PUY, photgraphed in Telford.
Philip Lamb

Purchased for operation on Shropshire Bus tendered routes were three Dennis Darts, all with Northern Counties bodywork. Two supplied in 1992 have the original version of the Paladin body as shown by 6, K6GOW as it loads passengers for Bridgnorth at Kidderminster. In 1995 a third Dart was delivered and this carries the latest styling of the Paladin and, as is the case with all three Darts, is fitted with high-back seating. Pictured in Bridgnorth during the summer of 1996 is 9, M409PUY. *Richard Godfrey*

GRAYLINE

Hartwool Ltd, Station Approach, Bicester, Oxfordshire, OX6 7BZ

NRU308M	Bristol VRT/SL2/6LX	Eastern Coach Works	H44/31F	1974	Ex Buckinghamshire Road Car, 1995
JOI4693	Bedford YMT	Plaxton Supreme III	C53F	1977	Ex Gray, Compton, 1982
KIB7026	Volvo B58-61	Duple Dominant II	C53F	1979	Ex Barratt, Nantwich, 1995
KIB7027	Leyland Leopard PSU5C/4R	Duple Dominant II	C53F	1979	Ex Prentice & McQuillan, Swanley, 1991
WWL537T	Bedford YMT	Plaxton Supreme III	C53F	1979	Ex Shaw, Maxey, 1980
EOI4376	Leyland Leopard PSU3F/4R	Plaxton Supreme IV Express	C53F	1981	Ex United, 1993
JND258V	Leyland Leopard PSU5C/4R	Duple Dominant II	C53F	1980	Ex University Bus, Hatfield, 1995
MJI1677	DAF SB2300DHS585	Berkhof Esprite 340	C53F	1985	Ex Limebourne, Battersea, 1988
MJI1679	DAF SB2300DHS585	Berkhof Esprite 340	C53F	1985	Ex Limebourne, Battersea, 1988
E215PWY	Volkswagen LT55	Optare City Pacer	B25F	1987	Ex Coach House Travel, Dorchester, 1995
MJI1676	Bova FHD12.290	Bova Futura	C49F	1988	Ex Embling, Guyhirn, 1992
E61SUH	Volkswagen LT55	Optare City Pacer	B25F	1988	Ex Rees & Williams, Morriston, 1995
E68SUH	Volkswagen LT55	Optare City Pacer	B25F	1988	Ex Rees & Williams, Morriston, 1995
449BHU	Bova FHD12.290	Bova Futura	C55F	1989	Ex Majestic, Shareshill, 1992
SPV860	Bova FHD12.290	Bova Futura	C55F	1990	
947CBK	MAN 10.180	Caetano Algarve	C31F	1990	
940HFJ	Scania K113CRB	Plaxton Paramount 3500 III	C53F	1990	Ex Moore, Sleaford, 1991
98CLJ	DAF SB2305DHS585	Caetano Algarve	C53F	1990	
KBC6V	Bedford YMT	Duple Dominant II	C53F	1980	Ex APT, Rayleigh, 1996
K105UFP	Dennis Javelin 12SDA2101	Caetano Algarve II	C53F	1993	Ex Euroline, Radford, 1995
L348MKU	Plaxton 425	Lorraine	C53F	1994	
N681AHL	Scania K93CRB	Berkhof Excellence 1000L	C55F	1995	

Previous Registrations:

449BHU	F690ONR	JOI4693	PPG2R	MJI1677	B682BTW
940HFJ	G566WVL	KIB7026	APH516T, MUD490, NLG734T	MJI1679	B684BTW
947CBK	G442WFC	KIB7027	EAP937V	SPV860	G441WFC
98CLJ	H322GBW	MJI1676	F109YFL	WWL537T	XEW321T, 947CBK
EOI4376	NDC242W				

Photographed heading towards Bicester in Oxfordshire is Graylines Volkswagen LT55, E215PWY. This Optare City Pacer is one of three of the type operated principally on Oxfordshire services.
Malc McDonald

The changes in coach design over the years is demonstrated in these two views. The upper picture, taken in New Road Oxford, shows KIB7026 a Leyland Leopard with the 12-metre PSU5 chassis and a Duple Dominant II body as supplied to National Bus Company. The lower picture shows a Plaxton Paramount-bodied Scania K113 which was photographed entering the coach park at Cheltenham Race Course in March 1996. *Richard Godfrey/David Donati*

BERLIN CITY TOUR
GUiDE FRIDAY
GUiDE FRIDAY
BERLIN CITY TOUR
GUIDE FRIDAY GmbH
COMMENTARY
KOMMENTAR
M·A·N
B AR 5461

THE EDINBURGH TOUR
EVERY 15 MINS EVERY DAY
WITH YOUR OWN GUIDE
NO NEED TO BOOK
JUST JUMP ON
THE EDINBURGH TOUR
EVERY 15 MINS EVERY DAY
OTO 585H
GUIDE FRIDAY LTD
TOWN AND CITY TOURS
THROUGHOUT GREAT BRITAIN

GUIDE FRIDAY

Guide Friday Ltd, Civic Hall, 14 Rother Street, Stratford-upon-Avon,
Warwickshire, CV37 6LU

Operations: Birmingham (Bir); Brighton (Bri); Cambridge (Cam); Cork; Dublin; Dundee; Edinburgh (Ed); Edinburgh Airport (EdA); Galway; Glasgow (Gla); Norwich; Oxford; Paris; Plymouth (Ply); Portsmouth (Por); Southampton (Sou); Stratford (Str); Windsor: York

Bir	OTO552M	Leyland Atlantean AN68/1R	East Lancashire	O47/33F	1974	Ex Nottingham, 1992
Bir	OTO582M	Leyland Atlantean AN68/1R	East Lancashire	O47/31F	1974	Ex Nottingham, 1992
Bri	WJY759	Leyland Atlantean PDR1/1	Metro Cammell	O44/33F	1963	Ex Plymouth, 1991
Bri	ERV254D	Leyland Atlantean PDR1/1	Metro Cammell	O43/33F	1966	Ex Southdown, 1991
Bri	GYS896D	Leyland Atlantean PDR1/1	Alexander A	O44/34F	1966	Ex Windsorian, Windsor, 1990
Bri	MLH304L	Daimler Fleetline CRG6LXB	MCW	O44/27F	1973	Ex Windsorian, Windsor, 1990
Bri	GTO334N	Leyland Atlantean AN68/1R	East Lancashire	O47/34F	1975	Ex Nottingham, 1990
Cam	MNU191P	Daimler Fleetline CRG6LX	Northern Counties	O47/30D	1976	Ex Nottingham, 1988
Cam	CWF733T	Leyland Atlantean AN68A/1R	Roe	O45/29D	1979	Ex South Yorkshire, 1991
Cam	CWF736T	Leyland Atlantean AN68A/1R	Roe	O45/31F	1979	Ex South Yorkshire, 1991
Cam	CWG744V	Leyland Atlantean AN68A/1R	Roe	O45/29D	1979	Ex South Yorkshire, 1991
Cam	CWG763V	Leyland Atlantean AN68A/1R	Roe	O45/31F	1979	Ex South Yorkshire, 1991
Cam	DWJ564V	Leyland Atlantean AN68A/1R	Roe	O45/31F	1979	Ex South Yorkshire, 1991
Cardiff	KHC814K	Leyland Atlantean PDR1A/1	East Lancashire	O45/31F	1972	Ex Eastbourne, 1989
Cardiff	KHC815K	Leyland Atlantean PDR1A/1	East Lancashire	O45/31F	1972	Ex Eastbourne, 1989
Dundee	UKV479R	Daimler Fleetline CRG6LX	Northern Counties	O47/31F	1976	Ex Nottingham, 1989
Ed	OTO573M	Leyland Atlantean AN68/1R	East Lancashire	O47/31F	1974	Ex Nottingham, 1989
Ed	OTO574M	Leyland Atlantean AN68/1R	East Lancashire	O47/31F	1974	Ex Nottingham, 1989
Ed	OTO584M	Leyland Atlantean AN68/1R	East Lancashire	O47/31F	1974	Ex Nottingham, 1989
Ed	OTO585M	Leyland Atlantean AN68/1R	East Lancashire	O47/31F	1974	Ex Nottingham, 1989
Ed	GRC887N	Leyland Atlantean AN68/1R	East Lancashire	O47/31F	1975	Ex Nottingham, 1989
Ed	GRC888N	Leyland Atlantean AN68/1R	East Lancashire	O47/32F	1975	Ex Nottingham, 1989
Ed	UKV473R	Daimler Fleetline CRG6LX	Northern Counties	O47/31F	1976	Ex Nottingham, 1989
Ed	UKV482R	Daimler Fleetline CRG6LX	Northern Counties	O47/31F	1976	Ex Nottingham, 1989
Ed	UKV470R	Daimler Fleetline CRG6LX	Northern Counties	O47/31F	1976	Ex Nottingham, 1989
Ed	ATV673T	Leyland Atlantean AN68A/1R	Northern Counties	H47/31D	1978	Ex Nottingham, 1996
EdA	GRC889N	Leyland Atlantean AN68/1R	East Lancashire	H47/32F	1975	Ex Nottingham, 1989
EdA	XNN664S	Leyland Atlantean AN68A/1R	Northern Counties	H47/31D	1978	Ex Nottingham, 1996
EdA	ARC638T	Leyland Atlantean AN68A/1R	East Lancashire	H47/33D	1978	Ex Nottingham, 1995
EdA	ARC639T	Leyland Atlantean AN68A/1R	East Lancashire	H47/33D	1978	Ex Nottingham, 1995
EdA	ARC642T	Leyland Atlantean AN68A/1R	East Lancashire	H47/33D	1978	Ex Nottingham, 1995
EdA	BTV650T	Leyland Atlantean AN68A/1R	East Lancashire	H47/33D	1978	Ex Nottingham, 1995
Gla	OTO571M	Leyland Atlantean AN68/1R	East Lancashire	H47/33F	1974	Ex Nottingham, 1989

CHL772 is a single-deck Daimler CVD6 and one of two of this type operated by Guide Friday at its home base, Stratford-upon-Avon. Guide Friday has become the major player in the tourist market and, as well as operating their own vehicles use other operators' vehicles in some places such as Lincoln, Chester, Bath and Llandudno.
Philip Lamb

Gla	ARC637T	Leyland Atlantean AN68A/1R	East Lancashire	H47/33D	1978	Ex Nottingham, 1996
Norwich	KSU839P	Leyland Atlantean AN68/1R	Alexander	H45/31F	1975	Ex Windsorian, Windsor, 1990
Oxford	264ERY	Leyland Titan PD3A/1	Park Royal	O43/33R	1963	Ex Leicester, 1978
Oxford	KHC817K	Leyland Atlantean PDR1A/1	East Lancashire	O45/31F	1972	Ex Eastbourne, 1988
Oxford	OTO543M	Leyland Atlantean AN68/1R	East Lancashire	H47/29D	1974	Ex Nottingham, 1989
Oxford	OTO549M	Leyland Atlantean AN68/1R	East Lancashire	H47/32F	1974	Ex Nottingham, 1990
Oxford	JAL876N	Leyland Atlantean AN68/1R	East Lancashire	O47/33D	1975	Ex Nottingham, 1992
Oxford	MPT314P	Leyland Atlantean AN68/1R	Eastern Coach Works	O45/27D	1975	Ex Oxford Bus Company, 1992
Oxford	UTV215S	Daimler Fleetline FE30AGR	Northern Counties	O47/33F	1978	Ex Leon's, Stafford, 1996
Oxford	UTV217S	Daimler Fleetline FE30AGR	Northern Counties	O47/33F	1978	Ex Nottingham, 1994
Paris	ARW938S	MAN SD200	Waggon Union	H53/35F	1978	Ex BVG, Berlin, 1995
Ply	ERV250D	Leyland Atlantean PDR1/1	Metro Cammell	O43/33F	1966	Ex Southdown, 1991
Ply	BTV652T	Leyland Atlantean AN68A/1R	East Lancashire	H47/33D	1978	Ex Nottingham, 1996
Ply	BTV657T	Leyland Atlantean AN68A/1R	East Lancashire	H47/33D	1978	Ex Nottingham, 1996
Por	XKO54A	Leyland Atlantean PDR1/1	Weymann	O44/33F	1963	Ex Robson Thornaby, 1992
Por	ETO161L	Daimler Fleetline CRG6LX	Willowbrook	O47/32F	1973	Ex Nottingham, 1985
Sou	WOW529J	Leyland Atlantean PDR1A/1	East Lancashire	O43/31F	1972	Ex Southampton, 1995
Str	HKL826	AEC Regal 0662	Beadle	OB35F	1946	Ex Hastings & District, 1989
Str	HKL836	AEC Regal 0662	Beadle	OB35F	1946	Ex Hastings & District, 1989
Str	CHL772	Daimler CVD6	Willowbrook	DP35F	1950	Ex Hastings & District, 1989
Str	CRU184C	Daimler Fleetline CRG6LX	Weymann	O45/29F	1965	Ex London Transport, 1984
Str	PRH257G	Leyland Atlantean PDR1A/1	Roe	O44/31F	1966	Ex ?, 199.
Str	KHC813K	Leyland Atlantean PDR1A/1	East Lancashire	O45/31F	1972	Ex Eastbourne, 1988
Str	GVO715N	Leyland Atlantean AN68/1R(LPG)	East Lancashire	H47/33F	1975	Ex Nottingham, 1992
Str	MUA873P	Bristol VRT/SL3/6LX(LPG)	Eastern Coach Works	H43/31F	1975	Ex ?, 199.
Str	KOU794P	Bristol VRT/SL3/6LXB(LPG)	Eastern Coach Works	H39/31F	1976	Ex City Line, 1996
Str	XNN665S	Leyland Atlantean AN68A/1R	Northern Counties	H47/31D	1978	Ex Nottingham, 1996
Str	CWG743V	Leyland Atlantean AN68A/1R	Roe	O45/29F	1979	Ex South Yorkshire, 1991
Str	KUB667V	Leyland Leopard PSU3E/4R	Plaxton Supreme IV	C53F	1980	Ex Duff, Sutton on Forest, 1993
Str	GWV926V	Leyland Leopard PSU3E/4R	Plaxton Supreme IV	C53F	1980	Ex Duff, Sutton on Forest, 1993
Winsor	GTO333N	Leyland Atlantean AN68/1R	East Lancashire	O47/34F	1975	Ex Nottingham, 1990
Winsor	GRC890N	Leyland Atlantean AN68/1R	East Lancashire	O47/34F	1975	Ex Nottingham, 1990
Winsor	BTV649T	Leyland Atlantean AN68A/1R	East Lancashire	O47/33F	1978	Ex Nottingham, 1995
York	DWU839H	Bristol VRT/SL2/6LX	Eastern Coach Works	O39/31F	1969	Ex York City & District, 1990
York	FWT956J	Bristol VRT/SL2/6LX	Eastern Coach Works	O39/31F	1970	Ex York City & District, 1990
York	GHL191L	Bristol VRT/SL2/6LX	Eastern Coach Works	O39/31F	1973	Ex Cambus, 1993
York	WWR417S	Bristol VRT/SL3/6LXB	Eastern Coach Works	O43/31F	1977	Ex York City & District, 1990
York	WWR418S	Bristol VRT/SL3/6LXB	Eastern Coach Works	O43/31F	1977	Ex York City & District, 1989
York	WWR419S	Bristol VRT/SL3/6LXB	Eastern Coach Works	O43/31F	1977	Ex York City & District, 1989
York	WWR420S	Bristol VRT/SL3/6LXB	Eastern Coach Works	O43/31F	1977	Ex York City & District, 1990
York	BKE861T	Bristol VRT/SL3/6LXB	Eastern Coach Works	O43/31F	1979	Ex Hastings & District, 1989

Cork	75KE520	Leyland Atlantean AN68/1R	East Lancashire	O47/31D	1975	Ex Nottingham, 1992
Dublin	73KE513	Leyland Atlantean AN68/1R	East Lancashire	O47/31F	1974	Ex Nottingham, 1989
Dublin	75KE509	Leyland Atlantean AN68/1R(LPG)	East Lancashire	O47/30F	1974	Ex Nottingham, 1989
Dublin	JAL880N	Leyland Atlantean AN68/1R	East Lancashire	H47/31D	1975	Ex Nottingham, 1993
Dublin	78KE547	Leyland Atlantean AN68/1R(LPG)	Northern Counties	H47/31D	1978	Ex Nottingham, 1996
Dublin	78KE549	Leyland Atlantean AN68/1R(LPG)	Northern Counties	H47/31D	1978	Ex Nottingham, 1996
Galway	75KE517	Leyland Atlantean AN68/1R	East Lancashire	O47/33F	1975	Ex Nottingham, 1990
Galway	75KE521	Leyland Atlantean AN68/1R	East Lancashire	O47/34F	1975	Ex Nottingham, 1990

Seville Leyland Atlantean AN68/1R East Lancashire O47/29D 1974 Ex Nottingham, 1989-90

SE4063-BU SE4065-BU SE4066-BU SE0032-BX SE0033-BX
SE4064-BU

Berlin MAN SD200 Waggon Union O53/35D 1977 Ex BVG, Berlin, 1995

B-AR5461 B-DD3438 B-DD3728 B-DD8568 B-DV1684
B-AR7901

Previous Registrations:

73KE513	OTO542M	B-AR7901	B-V2896	SE4064-BU	OTO575M
74KE509	OTO578M	B-DD3438	B-V3095	SE4065-BU	GTO332N
75KE517	GVO719N	B-DD3728	B-V3068	SE4066-BU	GRC886N
75KE520	GVO714N	B-DD8568	B-V3118	UKV470R	PAU199R
75KE521	GVO716N	B-DV1684	B-V3092	UKV473R	PAU196R
78KE547	XNN662S	SE0032-BX	GVO718N	UKV479R	PAU198R
78KE549	XXN663S	SE0033-BX	OTO583M	UKV482R	PAU197R
ARW938S	B-V3074	SE4063-BU	OTO572M	XKO54A	620UKM
B-AR5461	B-V3079				

H & H MOTORS

R M Horlick & M P Harris, Station Approach, Broad Meadows Yard, Ross on Wye, Herefordshire HR9 7AQ

NWS903R	Leyland National 11351A/2R		B49F	1977	Ex Red & White, 1995
SGR133R	Leyland National 11351A/1R		B49F	1977	Ex Burman, Dordon, 1993
BUH209V	Leyland National 11351A/1R		B49F	1979	Ex Red & White, 1994
GPA624V	Bedford YLQ	Plaxton Supreme IV	C45F	1980	Ex Nicholls, St Weonards, 1995
TIB4921	DAF MB200DKFL600	Plaxton Paramount 3500	C53F	1983	Ex Copelands, Meir, 1993
C315OFL	Ford Transit 190	Dormobile	B16F	1986	Ex Red & White, 1993
H569GMO	Freight Rover Sherpa	Premier	M16	1990	Ex Van, 1995

Previous Registrations:
TIB4921 EFK135Y, MIB516, FFA545Y

The continuing retrenchment from marginal work such as county and school contracts by some of the larger operators still provide some openings for the smaller operator. Only established some three years, H & H Motors employ three Leyland Nationals. Now in a blue and black livery, BUH209V is seen outside the depot showing schools service 858. The operator is also contracted for journeys on the Ross to Gloucester section of the Hereford to Gloucester service. *Ralph Stevens*

Page 42, top: **Guide Friday, while based in the South Midlands, operates vehicles far and wide. The upper picture shows one of the MAN-Büssing chassis with Wagon Union SD200 bodywork. New to the BVG in Berlin it remains in that city with Guide Friday who have converted it to open-top. These vehicles have two staircases and are dual-doored.** *Bottom:* **In Britain guided tours are provided in most towns and cities popular with tourists. Seen in Edinburgh is OTO585M, one of many former Nottingham vehicles in the fleet. It is normal practice to allocate buses to towns for at least a full season as each bus has lettering and pictures appropriate for the location.**

HARDINGS

Tanners Croft Garage Ltd, 335 Crabbs Cross Road, Redditch, Worcestershire B97 5HL

Depots : Boycott Trading Estate, Droitwich and Crabbs Cross Road, Redditch

HCR233	AEC Reliance 6U3ZR	Plaxton Elite III	C53F	1973	
NAB250P	AEC Reliance 6U3ZR	Plaxton Supreme III Express	C53F	1975	
914GAT	AEC Reliance 6U3ZR	Plaxton Supreme III Express	C53F	1976	
941GAT	AEC Reliance 6U3ZR	Plaxton Supreme III	C53F	1977	
VOI5888	Volvo B58-56	Duple Dominant II	C53F	1979	Ex JBS Coaches, Bedford, 1995
CIW9129	DAF MB200DKTL600	Plaxton Supreme IV	C53F	1981	Ex Tellyn, Witham, 1989
954GAT	Volvo B10M-61	Caetano Alpha	C53F	1981	Ex Miller, Foxton, 1983
KWP111X	Volvo B10M-56	Plaxton Supreme V Express	C53F	1982	
HCR601	Van Hool T815	Van Hool	C30FT	1983	Ex Droitwich Executive, 1986
D470DWP	Ford Transit 160	Chassis Developments	M12	1986	Ex AMK, Passfield, 1994
DSK514	Van Hool T815	Van Hool	C53F	1987	Ex Bicknell, Godalming, 1989
DSK516	Van Hool T815	Van Hool	C53F	1987	Ex Bicknell, Godalming, 1989
DSK593	Mercedes-Benz L307D	Yeates	M12	1987	Ex Snell, Newton Abbot, 1989
D158BPH	Kässbohrer-Setra S215HD	Kässbohrer	C F	1987	Ex Loveridge, Marks Tey, 1995
DSK515	Scania K112CRS	Jonckheere Jubilee P50	C57F	1988	
A4HCR	Toyota Coaster HB31R	Caetano Optimo	C21F	1989	
DSK594	Mercedes-Benz L307D	Yeates	M12	1989	Ex Excelsior, Bournemouth, 1995
G185JWP	Volvo B10M-60	Jonckheere Deauville P599	C51FT	1990	
G186JWP	Volvo B10M-60	Jonckheere Deauville P599	C51FT	1990	
H2HCR	Scania K113CRB	Van Hool Alizée	C51FT	1991	
H3HCR	Scania K113CRB	Van Hool Alizée	C51FT	1991	
H4HCR	Scania K113CRB	Van Hool Alizée	C49FT	1991	Ex Shearings, 1994
H5HCR	Scania K113CRB	Van Hool Alizée	C49FT	1991	Ex Shearings, 1994
J2HCR	Scania K113CRB	Van Hool Alizée	C51FT	1992	
J3HCR	Scania K113CRB	Van Hool Alizée	C51FT	1992	
K2HCR	Scania K113CRB	Van Hool Alizée	C51FT	1993	
K3HCR	DAF MB230LT615	Van Hool Alizée	C F	1993	
K4HCR	Toyota Coaster HDB30R	Caetano Optimo II	C21F	1993	
L2HCR	Scania K113CRB	Irizar Century 12.35	C49FT	1994	
M2HCR	EOS E180Z	EOS90	C49FT	1995	
M3HCR	Scania K113CRB	Van Hool Alizée	C49FT	1995	
N2HCR	Scania K113CRB	Irizar Century 12.35	C49FT	1996	
N3HCR	Scania K113CRB	Irizar Century 12.35	C49FT	1996	

Previous Registrations:

914GAT	VNP154R
941GAT	YNP994R
954GAT	MEW151W, 954GAT, EUY641W
A4HCR	F861WWP
CIW9129	SPP609W
D470DWP	D514OPP, DSK594
DSK514	D78APC
DSK515	E440LNP
DSK516	D79APC
DSK593	D567TOD
DSK594	8998LJ, D877YRU, A20EXC, D95AEL
F352YAB	F431VUY, A3HCR
H4HCR	H153DVM
H5HCR	H151DVM
HCR233	VWP457M
HCR601	BRW737Y, TSV906, CIW9129
VOI5888	RVY937T

Tanners Croft Garage Ltd was formed from a long-established partnership between Messrs Hardy and Dyson which grew with the expansion of the "new town" of Redditch. One of the first heavy-weight purchases was AEC Reliance NAB250P, seen *(opposite top)* in Worcester. In 1983 the business of Everton Goldliner of Droitwich was acquired, bringing with it a number of continental coaches which had been mostly absent from the Hardy and Dyson fleet. Among them were integrally and semi-integrally constructed vehicles and these have continued to be popular. Indeed, only the AEC's, Volvo's and one DAF in the fleet feature traditional construction. First of three Irzar-bodied Scanias which are semi-integrals, was L2HCR, seen here in Oxford.
Ralph Stevens/Ian Kirkby-David Donati Collection

HARDINGS
HARDINGS
HARDINGS
NAB 250P

HARDINGS
GOLDLINER EMPEROR 1
HARDINGS
SCANIA
L2 HCR

HEYFORDIAN

Heyfordian Travel Ltd, Orchard Lane, Upper Heyford, Oxfordshire, OX5 3LB

Depots : Rabans Lane Ind Est, Aylesbury; Harefield Marine, Harefield; Bellfield Road, High Wycombe; Lamarsh Road, Oxford, Orchard Lane, Upper Heyford and Downs Road, Witney.

FIL7662	Leyland Leopard PSU3A/4R	Plaxton Elite	C52F	1970	
1636VB	AEC Reliance 6U2ZR	Plaxton Supreme III	C53F	1976	
FIL8446	Bedford YRQ	Plaxton Supreme III	C45F	1976	
FIL8317	Bedford YMT	Plaxton Supreme III Express	C53F	1976	Ex Hills of Tredegar, 1982
FIL8441	Bedford YMT	Plaxton Supreme III Express	C53F	1976	Ex Hills of Tredegar, 1982
FBZ7356	Bedford YMT	Plaxton Supreme III Express	C53F	1976	Ex Hills of Tredegar, 1982
4128AP	AEC Reliance 6U2R	Plaxton Supreme III	C53F	1977	
1430PP	AEC Reliance 6U2R	Plaxton Supreme III	C53F	1977	
7209RU	AEC Reliance 6U2R	Plaxton Supreme III	C53F	1977	
3150MC	AEC Reliance 6U2R	Plaxton Supreme III	C53F	1977	
DFC884R	Bedford YMT	Plaxton Supreme III	C53F	1977	Ex Frostway, Upper Heyford, 1987
7298RU	Bedford YMT	Plaxton Supreme III	C53F	1978	Ex Premier, Watford, 1987
7223MY	Bedford YMT	Plaxton Supreme III	C53F	1978	Ex Smith, Rickmansworth, 1987
7396LJ	Bedford YMT	Plaxton Supreme III	C53F	1979	Ex Frostway, Upper Heyford, 1987
7845LJ	AEC Reliance 6U2R	Duple Dominant II Express	C53F	1979	Ex London Country, 1985
7958NU	AEC Reliance 6U2R	Duple Dominant II Express	C53F	1979	Ex London Country, 1985
8252MX	AEC Reliance 6U2R	Duple Dominant II Express	C53F	1979	Ex London Country, 1985
8779KV	AEC Reliance 6U2R	Duple Dominant II Express	C53F	1979	Ex London Country, 1985
HSV720	Bedford YMT	Plaxton Supreme IV	C53F	1979	Ex Premier, Watford, 1987
481HYE	Bedford YMT	Plaxton Supreme IV	C53F	1979	Ex Premier, Watford, 1987
9197WF	Bristol VRT/SL3/6LXB	Alexander AL	H44/31F	1979	Ex Cardiff Bus, 1994
7034KW	Bristol VRT/SL3/6LXB	Alexander AL	H44/31F	1979	Ex Cardiff Bus, 1994
3762KX	Bristol VRT/SL3/6LXB	Alexander AL	H44/31F	1980	Ex Cardiff Bus, 1994
9945NE	Bristol VRT/SL3/6LXB	Alexander AL	H44/31F	1980	Ex Cardiff Bus, 1994
VSF438	Bedford YMT	Plaxton Supreme IV	C53F	1981	Ex Smith, Rickmansworth, 1987
748ECR	Bedford YMT	Plaxton Supreme IV	C53F	1981	Ex Smith, Rickmansworth, 1987
943YKN	Bedford YMT	Plaxton Supreme IV	C53F	1981	Ex Smith, Rickmansworth, 1987
FBZ7357	Bova EL26/581	Bova Europa	C53F	1981	Ex Alder Valley South, 1991
3078RA	Bova EL26/581	Bova Europa	C53F	1982	Ex Staines Crusader, Clacton, 1984
4068MH	Bova EL26/581	Bova Europa	C53F	1982	Ex The Londoners, Nunhead, 1987
6940MD	Bova EL26/581	Bova Europa	C53F	1982	Ex The Londoners, Nunhead, 1987
3139KV	Bova EL26/581	Bova Europa	C53F	1983	Ex Wallace Arnold, 1987
6230NU	Bova EL26/581	Bova Europa	C53F	1983	Ex Wallace Arnold, 1987
2482NX	Bova EL26/581	Bova Europa	C53F	1983	Ex Wallace Arnold, 1987

Five AEC Reliance coaches with Duple Dominant II bodies originally with London Country are used by Heyfordian on school contract work until being sold during the summer of 1996. These now carry Heritage index marks. Seen parked at Aylesbury is 2462FD, originally YPL67T.
Ralph Stevens

Photographed entering Trafalgar Square in London, Heyfordian HIL2295 is a Scania K112 with Jonckheere Jubilee P599 bodywork which features a lowered driving position providing some of the benefits of a higher-floor coach without the attendant weight penalty. *Colin Lloyd*

2705TD	Bova EL26/581	Bova Europa	C40F	1983	Ex Eastern National, 1987
PVV316	Bova EL26/581	Bova Europa	C52F	1983	Ex Tourmaster, Dunstable, 1987
5057VC	Bova EL26/581	Bova Europa	C53F	1983	
5701DP	Bova EL26/581	Bova Europa	C53F	1983	
2779UE	Bova EL26/581	Bova Europa	C52F	1983	Ex Grayline, Bicester, 1984
4827WD	Scania K112CRS	Jonckheere Jubilee P599	C51FT	1984	Ex BTS Borehamwood, 1991
HIL2295	Scania K112CRS	Jonckheere Jubilee P599	C49FT	1984	Ex Goodwin, Stockport, 1992
ESU940	Scania K112CRS	Jonckheere Jubilee P599	C51FT	1984	Ex Goodwin, Stockport, 1994
868AVO	Scania K112CRS	Jonckheere Jubilee P599	C49FT	1984	Ex Hardings, Huyton, 1992
1264LG	Scania K112CRS	Jonckheere Jubilee P50	C53F	1985	
5089LG	Scania K112CRS	Jonckheere Jubilee P50	C53F	1985	
XCT550	Scania K112CRS	Jonckheere Jubilee P599	C51FT	1985	Ex Cross Gates Coaches, 1992
6960TU	Scania K112CRS	Jonckheere Jubilee P599	C57F	1985	Ex Cresswell, Moira, 1991
2185NU	Bova EL29/581	Bova Europa	C53F	1985	Ex Main Line, Tonyrefail, 1987
8216FN	DAF SB2300DHS585	Plaxton Paramount 3200 II	C53F	1985	
SJI4428	Scania K112CRS	Jonckheere Jubilee P599	C51FT	1985	Ex Constable, Long Melford, 1995
2110UK	Leyland Tiger TRCTL11/3RH	Duple 340	C51F	1986	Ex Crosville Wales, 1995
SJI5861	Leyland Tiger TRCTL11/3RH	Duple 340	C49FT	1987	Ex Trelawney Tours, Hayle, 1995
LDZ2502	Scania K112CRS	Jonckheere Jubilee P599	C49FT	1987	Ex Gillespie, Kelty, 1992
LDZ2503	Scania K112CRS	Jonckheere Jubilee P599	C51FT	1987	Ex Buddens, Romsey, 1992
9769UK	Volvo B10M-61	Duple 340	C53F	1988	Ex Westbus, Ashford, 1993
8548VF	Volvo B10M-61	Duple 340	C53F	1988	Ex Westbus, Ashford, 1993
9682FH	Volvo B10M-61	Duple 340	C53F	1988	Ex Westbus, Ashford, 1993
FIL7664	Volvo B10M-61	Duple 340	C53F	1988	Ex Westbus, Ashford, 1995
6595KV	Aüwaerter Neoplan N122/3	Aüwaerter Skyliner	CH57/20DT	1989	Ex Voyager, Selby, 1992
2622NU	Toyota Coaster HB31R	Caetano Optimo	C21F	1990	
1435VZ	Hestair Duple SDA1512	Duple 425	C57F	1990	Ex Limebourne, Battersea, 1994
4078NU	Auwaerter Neoplan N122/3	Auwaerter Skyliner	CH57/22CT	1991	Ex Mayo, Caterham, 1995
YAY537	Volvo B10M-60	Van Hool Alizée	C49FT	1992	Ex Durham City Coaches, 1996
J670LGA	Volvo B10M-60	Van Hool Alizée	C49DT	1992	Ex Mackie's of Alloa, 1996
J687LGA	Volvo B10M-60	Van Hool Alizée	C49DT	1992	Ex Priory Coaches, Gosport, 1996
J689LGA	Volvo B10M-60	Van Hool Alizée	C49DT	1992	Ex Bee Line Buzz, 1996

Two tri-axle Aüwaerter Neoplan N122s operate with Heyfordian and these carry the Aüwaerter Skyliner double-deck coach body as on this particular vehicle. Photographed in Oxford, 6595KV illustrates the dual doorway arrangement, with access to the rear luggage area also visible.
Robert Edworthy

L535XUT	Toyota Coaster HZB50R	Caetano Optimo III	C18F	1994	
L26CAY	MAN 10-190	Caetano Algarve II	C33FT	1994	
L740YGE	Volvo B10M-62	Jonckheere Deauville 45	C49FT	1994	Ex Park's, 1995
L743YGE	Volvo B10M-62	Jonckheere Deauville 45	C49FT	1994	Ex Park's, 1995
L745YGE	Volvo B10M-62	Jonckheere Deauville 45	C49FT	1994	Ex Park's, 1996

Previous Registrations:

1264LG	B157YBW	6595KV	F625OWJ, NIW2235	FBZ7356	MHB854P, 4827WD, UFC144P
1430PP	XWL801R	6940MD	YMV351Y	FBZ7357	KEP640X
1435VZ	G648YVS	6960TU	B71MLT, C47CKR	FIL7297	SUD464P
1636VB	PUD371P	7034KW	WTG379T	FIL7661	-
2110UK	C72KLG	7209RU	XWL798R	FIL7662	VUD384H
2185NU	B246YKX	7223MY	CMJ99T	FIL7663	-
2482NX	FUA398Y	7298RU	CMJ3T	FIL7664	E171OMU
2622NU	G152ELJ	7396LJ	YOG965T	FIL8317	MHB852P
2705TD	BGX649Y	748ECR	UNK101W	FIL8441	MHB853P
2779UE	FWL782Y	7845LJ	YPL84T	FIL8446	SUD465P
3078RA	DEV807X	7958NU	YPL73T	HIL2295	A131XNH
3139KV	FUA395Y	8216FN	From New	HSV720	KPP8V
3150MC	XWL799R	8252MK	YPL61T	J670LGA	J456HDS, LSK496
3762KX	CTX387V	8548VF	E174OMU	J687LGA	J457HDS, LSK497
4068MH	JAB311X	868AVO	A52JLW	J689LGA	J460HDS, LSK500
4078NU	H297GKN	8779KV	YPL77T	LDZ2502	D313VVV
4128AP	XWL800R	9197WF	WTG364T	LDZ2503	D312VVV
481HYE	KPP9V	943YKN	UNK102W	PVV316	JRO615Y
4827WD	A59JLW, ESU930, A545TMJ	9467MU	-	SJI4428	B505CBD, RDU4, B989MAB
5057VC	From New	9682FH	E1750MU	SJI5861	D272FAS, SJI1998
5089LG	B156YBW	9769UK	E173OMU	VSF438	KPP100V
5701DP	From New	9945NE	CTX390V	XCT550	B504CBD, HYY3, B984MAB
6230NU	FUA396Y	ESU490	A60JLW	YAY537	F483OFT

Livery: Off-white, orange, red and yellow

HOLLANDS

G A Mole, Crossley Estate, Mill Street, Kidderminster, Worcestershire DY11 3XG

OHA436W	Ford Transit 190	Reeve Burgess Reebur	C17F	1981	Ex Mini-Mac, Kippax, 1989
KGE299Y	Ford Transit 190	Dormobile	M16	1983	Ex Atkin & Dickenson, Dringhouses, 1989
A607HNF	Ford Transit 190	Mellor	M16	1983	Ex Woodstones, Kidderminster, 1994
C918YBF	Ford Transit 190	Deansgate	M16	1986	Ex Private Owner, 1995
C380KUX	Ford Transit 160	?	M14	1986	Ex Private Owner, 1995
D392KND	Ford Transit VE6	Mellor	M16	1986	Ex Owen, Kidderminster, 1991
D723JUB	Freight Rover Sherpa	Carlyle	B16F	1986	Ex Horrocks, Brockton, 1993
D247OOJ	Freight Rover Sherpa	Carlyle	B20F	1987	Ex Victoria Shuttle, London, 1990
D133NON	Freight Rover Sherpa	Carlyle	B20F	1987	Ex The Wright Company, Wrexham, 1991
D118TFT	Freight Rover Sherpa	Carlyle	B18F	1987	Ex Merry Hill Minibus, 1991
F56SAD	Renault Master T35D	Coachwork Walker	M14L	1988	Ex Manchester College of Art, 1995
K436AAV	Ford Transit VE6	Ford	M14	1993	Ex Gilbert Rice, Cambridge, 1996
M433PUY	Ford Transit VE6	Ford	M14	1995	
M396SAB	Ford Transit VE6	Ford	M14	1995	

Livery: White and blue

Hollands of Kidderminster operate tendered services for Worcestershire, especially schools work. Seen in its home town is the only Renault in the fleet, F56SAD. The Master T35 product is fitted with a rear door and wheelchair lift and is particularly successful in a welfare role as its low-floor arrangement makes access easier and often a simple ramp at the rear can avoid the fitting of a bulkier and heavy tail lift. *Richard Godfrey*

JBC / MALVERNIAN

G G & M A Crump, Newton Road, Malvern Link, Worcestershire WR14

17	PKG108R	Bedford YMT	Plaxton Supreme III	C53F	1976	Ex Jones Brothers, Malvern, 1982
19	KFO570P	Bedford YRT	Duple Dominant	C53F	1975	Ex Jones Brothers, Malvern, 1982
21	AUJ747T	Bedford YLQ	Duple Dominant II	C45F	1979	Ex Jones Brothers, Malvern, 1982
23	LFB681P	Leyland Leopard PSU3C/4R	Plaxton Supreme III Express	C53F	1975	Ex Brian Isaac, Swansea, 1990
	Q305VAB	Leyland Leopard PSU5B/4R	Duple Dominant II	C53F	1979	Ex Titterington, Blencow, 1991
24	FFJ473V	Bedford YMT	Plaxton Supreme IV	C53F	1979	Ex Tidworth Coaches, 1987
	DFB222W	Volvo B58-56	Plaxton Supreme IV	C53F	1981	Ex Berkeley, Paulton, 1995
28	OIB9385	Bova EL26/581	Duple Calypso	C53F	1984	Ex Moseley demonstrator, 1984
29	OIB9386	LAG G350Z	LAG Panoramic	C49FT	1984	Ex Marchwood, Totton, 1989
	SDR450T	Volvo B58-61	Duple Dominant II	C51F	1979	Ex Prospect, Lye, 1996
30	OIB9387	Dennis Javelin 12SDA1907	Duple 320	C53F	1989	Ex Antler, Rugeley, 1989
33	179EJU	Leyland Tiger TRCTL11/3RH	Plaxton Paramount 3500 II	C48FT	1985	Ex Owens, Knighton, 1994

Previous Registrations:

179EJU	From new	OIB9386	A270KEL	Q305VAB	SYU713S, BIB8281
OIB9385	A527AUY	OIB9387	F260OFP		

Livery: Beige, orange and brown

Jones Brothers Coaches was founded in 1947 and ran from Upton-on-Severn for fifteen years before moving to Malvern Link in 1963. In 1973 Mr H.S. Jones became sole proprietor and in 1982 the business passed to the Crump family and the Malvernian name introduced with Jones Brothers gradually being represented by the JBC logo. OIB9387 is the only Dennis Javelin in the fleet, and was purchased almost new from Antler of Rugeley. *Robert Edworthy*

JAMES BEVAN

James Bevan (Lydney) Ltd, Bus Station, Hams Road, Lydney GL15 5PE

DUP143S	Leyland Leopard PSU3E/4R	Plaxton Supreme III	C53F	1978	Ex Carney's Coaches, Sunderland, 1979
BEP967V	Bristol VRT/SL3/501	Eastern Coach Works	H44/31F	1979	Ex Brewers, 1994
ECY987V	Bristol VRT/SL3/501	Eastern Coach Works	H43/31F	1980	Ex SWT, 1994
EBZ8205	Auwaerter Neoplan N122/3	Auwaerter Skyliner	CH57/20CT	1987	Ex Trathens, 1995
E735VWJ	Mercedes-Benz 609D	Whittaker Europa	C24F	1987	Ex Vantage Coach Hire, Romford, 1991
E323UUB	Volvo B10M-61	Plaxton Paramount 3500 III	C53F	1988	Ex Wallace Arnold, 1991
F886SRT	DAF SB2305DHTD585	Plaxton Paramount 3200 III	C53F	1989	Ex Galloway, Mendlesham, 1993
H64XBD	MAN 16.290	Jonckheere Deauville P599	C51FT	1991	Ex Aztec, Bristol, 1994

Previous Registrations:
EBZ8205 XWC18

Livery: White, red, orange and yellow: silver (Silver Jubilee livery) E323UUB

1992 was the year of James Bevan's silver jubilee and a coach was acquired and painted to celebrate the event. However, the business was founded on one started before the First World War by Mr H T Letheren. Now based in the former bus station in Lydney, several services are are operated along the north bank of the lower reaches of the River Severn. Two Leyland-engined Bristol VRTs and new to South Wales are operated and pictured in Lydney is BEP967V, with Eastern Coach Works.
Robert Edworthy

JOHNSONS

Johnsons (Henley) Ltd, 32 High Street, Henley-in-Arden, Warwickshire B95 5AN

Depot :Birmingham Road, Henley-in-Arden

GVO721N	Leyland Atlantean AN68/1R	East Lancashire	H47/33F	1975	Ex City of Nottingham, 1991
ARC646T	Leyland Atlantean AN68A/1R	East Lancashire	H47/33F	1978	Ex Guide Friday, Stratford, 1995
JKV430V	Ford R1114	Plaxton Supreme IV Express	C53F	1980	Ex Smith, Alcester, 1984
TXI6708	Ford R1114	Plaxton Supreme IV	C53F	1980	Ex Nash's, Smethwick, 1990
RNK327W	Ford R1014	Plaxton Supreme IV	C45F	1981	Ex Patterson, Birmingham, 1986
RUE353W	Ford R1114	Duple Dominant II	C53F	1981	Ex Hazeldine, Bilston, 1988
PNW321W	Ford R1114	Plaxton Supreme IV	C53F	1981	Ex Trailways, Four Oaks, 1990
TXI6709	Ford R1114	Plaxton Supreme IV	C53F	1981	Ex Nash's, Smethwick, 1990
TXI6704	Bova FHD12.280	Bova Futura	C49FT	1984	Ex Burton, Fellback, 1988
RJI8683	DAF MB200DKFL600	Van Hool Alizée	C48FT	1984	Ex Shaw, Warwick, 1993
TXI6705	Bova FHD12.280	Bova Futura	C53F	1985	Ex Wiffen, Finchingfield, 1989
TXI6710	Volvo B9M	Plaxton Paramount 3200 II	C36F	1985	Ex Pulham's, Bourton, 1991
TXI6706	Bova FHD12.280	Bova Futura	C49FT	1986	Ex Wiffen, Finchingfield, 1990
RJI8682	Bova FLD12.250	Bova Futura	C57F	1986	Ex Spanish Speaking Services, 1992
TXI6707	DAF MB200DKFL600	Duple	C53F	1986	Ex Fletcher, Studley, 1989
D974PJW	Freight Rover Sherpa	Carlyle	B18F	1987	Ex Shaw, Warwick, 1993

Johnsons' base is the attractive town of Henley-in-Arden, just off the M40 in Warwickshire. The only Volvo in the fleet is TXI6710, a 10-metre B9m with Plaxton Paramount 3200 bodywork. It was pictured near Waterloo rail station in London. *Colin Lloyd*

Newer vehicles with Johnsons have included several Bova Futura integral coaches from The Netherlands. These are now imported through Optare, following its earlier association with the DAF group. The type now numbers fourteen in the fleet, including the Club variant which is a lower specification, competitively priced vehicle. Seen at Dorridge during 1996 is TXI6704. *David Cole*

RJI8689	Mercedes-Benz 609D	Whittaker Europa	C21F	1988	Ex Execubus, Booker Common, 1991
RJI8687	Dennis Javelin 12SDA1907	Plaxton Paramount 3200 III	C53F	1988	Ex Wainfleet, Nuneaton, 1994
RJI8684	Bova FHD12.290	Bova Futura	C53F	1989	Ex Stock, Kemsing, 1991
RJI8681	Bova FHD12.290	Bova Futura	C49FT	1989	Ex Fishwick, Leyland, 1992
F892JHA	Bova FHD12.290	Bova Futura	C25FT	1989	Ex Flights, Birmingham, 1996
RJI8685	Bova FHD12.290	Bova Futura	C49FT	1989	Ex Fishwick, Leyland, 1992
RJI8686	Dennis Javelin 12SDA1907	Plaxton Paramount 3200 III	C53F	1989	Ex Shaw, Warwick, 1993
RJI8688	DAF SB2305DHTD585	Plaxton Paramount 3200 III	C53F	1989	Ex Sharrock & Hill, West Houghton, 1994
RJI8690	Mercedes-Benz 609D	Whittaker Europa	C24F	1989	
G171XDX	Bova FHD12.290	Bova Futura	C49FT	1989	Ex Galloway, Mendlesham, 1993
G420WFP	Bova FHD12.290	Bova Futura	C55F	1989	Ex Wray, Harrogate, 1993
H180EJF	Toyota Coaster HDB30R	Caetano Optimo II	C21F	1991	
G997OVA	Bova FHD12.290	Bova Futura	C53F	1991	Ex Fishpoole, Teversham, 1991
J13OVA	Bova FHD12.290	Bova Futura	C53F	1991	Ex Bennett, Chieveley, 1995
L549OWC	Ford Transit VE6	Ford	M8	1994	Ex Ford Motor Co, Dagenham, 1995
L584EPC	Ford Transit VE6	Ford	M8	1994	Ex Budget Rentals, 1996
M10JMJ	Bova FHD12.340	Bova Futura	C51FT	1994	
M10RGJ	Bova FHD12.340	Bova Futura	C51FT	1995	
M173SBT	Bova FLD12.270	Bova Futura Club	C53F	1995	Ex Bova demonstrator, 1995
N10JRJ	Bova FLD12.340	Bova Futura	C49FT	1996	

Previous Registrations:

F892JHA	F907CJW, 245DOC, 1FTO	RJI8686	F237OFP	TXI6705	B311VVG
RJI8681	F550YCW	RJI8687	F375MUT, MIW5792, F911VVC	TXI6706	C659KHK
RJI8682	C338VRY	RJI8688	G254EHD	TXI6707	D100MDS
RJI8683	A359JJU	RJI8689	E609AWA	TXI6708	FOP563V
RJI8684	F260NUT	RJI8690	F864FWB	TXI6709	KOH596W
RJI8685	F40BFR	TXI6704	A257SBM, BIL736, A622RUG	TXI6710	C361HGF

Livery: White and fawn

KEN ROSE

Ken Rose Coaches (Broadway) Ltd, c/o Spiers & Hartwell Ltd, Blackminster, Evesham, Worcestershire WR11 5YM

VUR896W	Volvo B58-61	Duple Dominant IV	C57F	1981	Ex Travel Line, Abbots Langley, 1994
UWA579Y	Ford R1114	Duple Dominant IV	C53F	1983	Ex Stevens, Bristol, 1992
D957WJH	Freight Rover Sherpa	Dormobile	B16F	1986	Ex Hampshire Bus, 1992
D869BDG	Ford Transit 190	Dormobile	M16	1987	Ex RCJ Hire, Winchcombe, 1992
F146UFH	Leyland-DAF 400	Leyland-DAF	M16	1989	Ex Brimm, Honeybourne, 1992
K441ATF	Mercedes-Benz 410D	Mercedes-Benz	M15	1993	Ex Van, 1996
M139LNP	Mercedes-Benz 814D	Plaxton Beaver	C33F	1994	

Previous Registrations:
VUR896W PDJ903W, FSU386

Livery: White

The business of Ken Rose employs an all-white livery and is shown in this picture of M139LNP, a Mercedes-Benz 814 with Plaxton Beaver bodywork. The swing door of the coach variant of the Beaver model is illustrated in this view taken at the depot. The business has moved from the attractive Cotswold village of Broadway since the previous edition of this Bus Handbook. Operations were started in the early 1970s with a single minibus and that type has predominated in the intervening years. *Robert Edworthy*

KESTREL

P J Coombe & M Wood, Dunhampton Garage, Dunhampton, Worcestershire DY13 9SW

	HIL4995	Ford R1115	Plaxton Paramount 3200	C53F	1983	Ex Hyth & Waterside, Hardley, 1988
	NIW7756	Bova EL28/581	Bova Europa	C49FT	1984	Ex Mawby, Ombersley, 1994
	B76SFO	Bedford YMP	Plaxton Paramount 3200	C31F	1984	Ex Edwards Bros, Tiers Cross, 1994
	C108HKG	Ford Transit	Robin Hood	B16F	1986	Ex Red & White, 1995
	D396LRL	Ford Transit	Steedrive	M15	1987	Ex Coombs, Weston-super-Mare, 1993
	D708WEY	Iveco Daily 49.10	Robin Hood City Nippy	DP16F	1987	Ex Roberts, Colwyn Bay, 1995
46	E200YTM	Ford Transit	Chassis Developments	M16	1987	
	E903VWG	Ford Transit	Coachcraft	M16	1987	
	G999KJX	DAF MB230LT615	Van Hool Alizée	C53F	1990	Ex Regina, Blaenau Ffestiniog, 1995

Previous Registrations:

D708WEY	D38MAG, A3WTR	HIL4995	BLJ719Y	NIW7756	A736HFP

Kestrel minicoaches was started in the 1970's and changed hands twice before coming under the control of the present partnership in the 1990's. For some time a feature of the fleet was the use of index marks containing 111, 222 etc. for new vehicles. Photographed in Kidderminster, Kestrel E200YTM shows both private charter and boards for service 10. This vehicle also carries fleet number 46, the last of a fleet numbering scheme that reached 50 but did not continue. *Robert Edworthy*

LEWIS'S

NE & DF Lewis, Central Garage, Pailton, Warwickshire, CV23 0QH

PHR583R	Ford R1114	Duple Dominant	C53F	1977	Ex Davis, Minchinhampton, 1988
KVC385V	Ford R1114	Plaxton Supreme IV Express	C53F	1980	
KVC386V	Ford R1114	Plaxton Supreme IV Express	C53F	1980	
OWK83W	Ford R1114	Plaxton Supreme IV Express	C53F	1980	
F366MUT	Dennis Javelin 11SDL1905	Plaxton Paramount 3200 III	C53F	1988	Ex Lester's, Long Wharton, 1995
F622SAY	Dennis Javelin 12SDA1916	Plaxton Paramount 3200 III	C53F	1989	
F623SAY	Dennis Javelin 12SDA1916	Plaxton Paramount 3200 III	C53F	1989	
F624SAY	Dennis Javelin 12SDA1916	Plaxton Paramount 3200 III	C49FT	1989	
G970WNR	Dennis Javelin 12SDA1916	Plaxton Paramount 3200 III	C53F	1989	
G422YAY	Dennis Javelin 12SDA1929	Plaxton Paramount 3200 III	C53F	1990	Ex Slack, Tansley, 1995
J201RAC	Dennis Javelin 12SDA1929	Plaxton Paramount 3200 III	C53F	1991	Ex Supreme, Coventry, 1994
J718KBC	Dennis Javelin 12SDA1929	Plaxton Paramount 3200 III	C51FT	1992	
J719KBC	Dennis Javelin 12SDA1929	Plaxton Paramount 3200 III	C53FT	1992	

Previous Registrations:
PHR583R UDU870R, 3134AD

Livery: White, grey and red

The fleet of Lewis's of Rugby consists of only two chassis types. Fords were purchased until 1980 and Dennis Javelins for the deliveries since 1987. In the intervening period DAF was favoured but none now remain. All the Javelins have Plaxton Paramount bodywork and are represented by G422YAY, seen on a visit to Westminster. *Colin Lloyd*

LUGG VALLEY

Miss D J Staples, 131 Etnam Street, Leominster, Herefordshire, HR6 8AF

Depots : Station Yard Ind Est, Leominster.

VEN416L	Bedford YRT	Duple Dominant	C53F	1973	Ex Elsworth, Blackpool, 1975
NCJ800M	Bedford YRQ	Duple Dominant Express	C45F	1973	Ex Yeomans, Hereford, 1978
PVJ300M	Bedford YRQ	Duple Dominant	C45F	1974	Ex Canyon, Hereford, 1980
DJS203	Bedford YRT	Duple Dominant	C53F	1976	Ex Waterhouse, Polegate, 1979
MHB855P	Bedford YLQ	Plaxton Supreme III Express	C45F	1976	Ex Hills of Tredegar, 1988
MCJ900P	Bedford YLQ	Plaxton Supreme III Express	C45F	1976	
OVJ700R	Bedford YLQ	Plaxton Supreme III Express	C45F	1976	
NUY312T	Bedford YLQ	Plaxton Supreme IV	C45F	1976	Ex Rover, Bromsgrove, 1989
BFO400V	Bedford YMT	Plaxton Supreme IV Express	C53F	1980	
E65EVJ	Bedford CF	Steedrive Parflo	M12	1987	
A3NPT	Bedford YNV Venturer	Caetano Algarve	C53F	1988	Ex Buchanan, Stretton Sugwas
E559UHS	Volvo B10M-61	Plaxton Paramount 3200 III	C53F	1988	Ex Park's, 1990
F992HGE	Volvo B10M-61	Plaxton Paramount 3200 III	C53F	1989	Ex Park's, 1991
G57RGG	Volvo B10M-60	Plaxton Paramount 3500 III	C49FT	1990	Ex Park's, 1996
H902AHS	Volvo B10M-60	Plaxton Paramount 3500 III	C53F	1991	Ex Wallace Arnold, 1994

Previous Registrations:

A3NPT	E754JAY	DJS203	NNK823P

Livery: Ivory and green.

Miss D.J. Staples started operating in 1959 on contract and private hire work but later took up some stage services which Primrose had given up. Current services are between Leominster and Hereford including the 292 direct service, also operated by Primrose and Midland Red West, and the 426/502 services which serve the thinly populated villages either side of the A49. After many years of operating Bedfords, Lugg Valley have, latterly, introduced the Volvo B10M to the fleet. Four now operate, all are fitted with Plaxton Paramount 3500 bodywork and including E559UHS, originally came from Park's of Hamilton. *Robert Edworthy*

McLEANS

McLeans Coaches & Taxis Ltd, Unit 5, Two Rivers Ind Est, Station Lane, Witney, Oxfordshire, OX8 6YD

C652XDF	Mercedes-Benz L608D	Alexander AM	B20F	1986	Ex Cheltemham & Gloucester, 1996
D888MJA	Renault-Dodge S56	Northern Counties	B24F	1986	Ex GMN, 1995
HIL6649	Mercedes-Benz 709D	Jubilee	C20F	1987	Ex Wallington, Great Rollright, 1995
F358GBW	Mercedes-Benz 609D	Reeve Burgess Beaver	C23F	1988	
F274JWL	Mercedes-Benz 811D	Reeve Burgess Beaver	C33F	1988	
F308LBW	Volvo B10M-60	Plaxton Paramount 3500 III	C49FT	1989	
F850NJO	Mercedes-Benz 811D	Reeve Burgess Beaver	C33F	1989	
G99XUD	Mercedes-Benz 0303	Mercedes-Benz	C49FT	1989	
G564LWX	Mercedes-Benz 0303	Plaxton Paramount 3500 III	C50FT	1990	Ex Wallace Arnold, 1993
H808RWJ	Scania K113CRB	Plaxton Paramount 3500 III	C53FT	1990	
UJI6312	Volvo B10M-60	Plaxton Paramount 3500 III	C53F	1991	Ex Park's, 1992
N383EAK	Volvo B10M-62	Plaxton Première 350	C50FT	1996	
N384EAK	Volvo B10M-62	Plaxton Première 350	C50FT	1996	

Previous Registrations:

HIL6649 D603XNH
UJI6312 H814AHS

Livery: White and red; white (Grand UK) UJI6312

The South Midlands county of Oxfordshire is served by many operators. McLeans use minibuses on services though the majority of the fleet are coaches. Pictured on some distance away from Oxford, in London, is Scania K113 H808RWJ. *Colin Lloyd*

MARCHANT'S

Marchants Coaches Ltd, 61 Clarence Street, Cheltenham, Gloucestershire, GL51 3LB

Depot: Prestbury Road, Cheltenham

CRO671K	AEC Reliance 6U3ZR	Plaxton Elite III	DP60F	1972	Ex Edwards, Gloucester, 1983
GDF650L	AEC Reliance 6U3ZR	Plaxton Elite III Express	C53F	1973	
YDD109S	AEC Reliance 6U3ZR	Plaxton Elite III Express	C53F	1978	
XAK902T	Bristol VRT/SL3/501	Eastern Coach Works	H43/31F	1978	Ex RoadCar, 1996
KTL25V	Bristol VRT/SL3/6LXB	Eastern Coach Works	H43/31F	1979	Ex RoadCar, 1996
KTL26V	Bristol VRT/SL3/6LXB	Eastern Coach Works	H43/31F	1979	Ex RoadCar, 1996
LGB855V	Volvo B58-56	Plaxton Supreme IV	C53F	1979	Ex Dorset Queen, East Chaldon, 1988
NMJ284V	Volvo B58-56	Plaxton Supreme IV	C53F	1980	Ex Galleon, Stepney, 1986
NMJ286V	Volvo B58-56	Plaxton Supreme IV	C53F	1980	Ex Galleon, Stepney, 1986
NFH528W	Volvo B58-56	Plaxton Supreme IV Express	C53F	1980	
NFH530W	Volvo B58-56	Plaxton Supreme IV Express	C53F	1980	
UBC464X	Volvo B10M-61	Plaxton Supreme V	C57F	1981	Ex Nash's, Smethwick, 1989
WDF998X	Volvo B10M-56	Plaxton Supreme V Express	C53F	1982	
WDF999X	Volvo B10M-56	Plaxton Supreme V Express	C53F	1982	
JEY124Y	Volvo B10M-61	Plaxton P'mount 3200 (1990)	C53F	1983	Ex Arvonia, Llanrug, 1990
A233MDD	Volvo B10M-56	Plaxton Paramount 3200E	C53F	1984	Ex David Field, Newent, 1994
A342LDG	Volvo B10M-61	Plaxton Paramount 3500	C51FT	1984	
C452CWR	Volvo B10M-61	Plaxton Paramount 3500 II	C49FT	1986	Ex NTE Coaches, 1987
C455CWR	Volvo B10M-61	Plaxton Paramount 3500 II	C49FT	1986	Ex NTE Coaches, 1987
F660RTL	Volvo B10M-60	Plaxton Paramount 3200 III	C53F	1989	Ex Appleby's, 1994
G448CDG	Volvo B10M-60	Plaxton Paramount 3500 III	C53F	1990	
G993DDF	Volvo B10M-60	Plaxton Paramount 3500 III	C51FT	1990	

Previous Registrations:

A233MDD	A733JAY, 6349D, A899YOV, A4DOF	JEY124Y	MSU593Y, VYB704
F660RTL	F287OFE, 5517RH	TVN585	FAC10Y

Livery: Grey and red

Marchant's commenced in 1949 and operate frequent local services from Cheltenham. Heavyweight coaches have been favoured since the take-over of Kearseys in 1968. Illustrated here is F660RTL, a Volvo with Plaxton Paramount 3200 bodywork. With the exception of a Duple-bodied Bedford which joined the fleet in 1995, all the coaches are of Plaxton manufacture with chassis by AEC or, subsequently, Volvo.
Robert Edworthy

MIDLAND RED WEST

Midland Red West Ltd, Heron Lodge, London Road, Worcester, WR5 2EW

Depots : Abbey Road, Evesham; Friar Street, Hereford; New Road, Kidderminster; Plymouth Road, Redditch and Padmore Street, Worcester. **Outstations:** Bull Ring bus station, Birmingham; Bishops Castle, Bridgnorth, Hopton Heath and Ludlow.

Part of FirstBus plc

201-237 Dennis Lance 11SDA3107 Plaxton Verde B54F 1994

201	L201AAB	**209**	L209AAB	**217**	L217AAB	**224**	L224AAB	**231**	L231AAB
202	L202AAB	**210**	L210AAB	**218**	L218AAB	**225**	L225AAB	**232**	L232AAB
203	L203AAB	**211**	L211AAB	**219**	L219AAB	**226**	L226AAB	**233**	L233AAB
204	L204AAB	**212**	L212AAB	**220**	L220AAB	**227**	L227AAB	**234**	L234AAB
205	L205AAB	**213**	L213AAB	**221**	L221AAB	**228**	L228AAB	**235**	L235AAB
206	L206AAB	**214**	L214AAB	**222**	L322AAB	**229**	L229AAB	**236**	L236AAB
207	L207AAB	**215**	L215AAB	**223**	L223AAB	**230**	L230AAB	**237**	L237AAB
208	L208AAB	**216**	L216AAB						

238-256 Dennis Lance 11SDA3113 Plaxton Verde B49F 1995

238	M238MRW	**242**	M242MRW	**246**	M246MRW	**250**	M250MRW	**254**	M254MRW
239	M239MRW	**243**	M243MRW	**247**	M247MRW	**251**	M251MRW	**255**	M255MRW
240	M240MRW	**244**	M244MRW	**248**	M248MRW	**252**	M252MRW	**256**	M256MRW
241	M241MRW	**245**	M245MRW	**249**	M249MRW	**253**	M253MRW		

301-313 Dennis Dart 9.8SDL3054 Plaxton Pointer DP36F 1995

301	N301XAB	**304**	N304XAB	**307**	N307XAB	**310**	N310XAB	**312**	N312XAB
302	N302XAB	**305**	N305XAB	**308**	N308XAB	**311**	N311XAB	**313**	N313XAB
303	N303XAB	**306**	N306XAB	**309**	N309XAB				

341	N341EUY	Dennis Dart SLF SFD212	Plaxton Pointer	DP33F	1996	
656	PUK656R	Leyland National 11351A/1R		B49F	1977	Ex Midland Red, 1981
657	SOA657S	Leyland National 11351A/1R		B49F	1977	Ex Midland Red, 1981
658	SOA658S	Leyland National 11351A/1R		B49F	1977	Ex Midland Red, 1981

722-752 Leyland National 11351A/1R B49F 1978-79 Ex Midland Red, 1981

722	WOC722T	**743**	XOV743T	**746**	XOV746T	**749**	XOV749T	**752**	XOV752T
723	WOC723T	**744**	XOV744T						

The number of coaches used by Midland Red West has diminished with several of the early Leyland Leopards and Tigers now used on service duties. Seen in Worcester is Tiger 1011, A895KCL, an example of Plaxton Paramount 3200 Express bodywork and was transferred from Midland Red Coaches in 1986.
Richard Godfrey

Badgerline introduced the Dennis Lance to Midland Red West who now operates fifty-six of the type, the highest capacity vehicles in the fleet - no double-decks being operated. The second batch carry very suitable index marks as illustrated by 251, M251MRW, seen passing through Worcester when heading for Great Malvern in May 1996. *Richard Godfrey*

755	AFJ755T	Leyland National 11351A/1R		B50F	1979	Ex Western National, 1989
756	AFJ756T	Leyland National 11351A/1R		B50F	1979	Ex Western National, 1990
758	XOV758T	Leyland National 11351A/1R		B49F	1979	Ex Midland Red North, 1986
770	BVP770V	Leyland National 11351A/1R		B49F	1979	Ex Midland Red, 1981
853	Q553UOC	Leyland Leopard PSU3F/4R	Plaxton P'mount 3200 (1984)	C49F	1982	Ex Midland Red Coaches, 1986
854	Q276UOC	Leyland Leopard PSU3E/4R	Plaxton P'mount 3200 (1983)	C49F	1980	Ex Midland Red, 1981
1001	FEH1Y	Leyland Tiger TRCTL11/3R	Plaxton Paramount 3500	C50FT	1983	

1002-1007	Leyland Tiger TRCTL11/3RH	Plaxton Paramount 3200 II	C50FT*	1985	*1006 is C43FT; 1007 is C39FT

1002	B102JAB	**1004**	B104JAB	**1005**	B105JAB	**1006**	B106JAB	**1007**	B107JAB
1003	B103JAB								

1008	LOA832X	Leyland Tiger TRCTL11/3R	Plaxton Supreme IV	C51F	1981	Ex Midland Red Coaches, 1986
1010	EAH890Y	Leyland Tiger TRCTL11/3R	Plaxton Paramount 3200 E	C53F	1983	Ex Midland Red Coaches, 1986
1011	A895KCL	Leyland Tiger TRCTL11/3R	Plaxton Paramount 3200 E	C53F	1983	Ex Midland Red Coaches, 1986
1012	A896KCL	Leyland Tiger TRCTL11/3R	Plaxton Paramount 3200 E	C53F	1983	Ex Midland Red Coaches, 1986
1013	A678KDV	Leyland Tiger TRCTL11/3R	Plaxton Paramount 3500	C48FT	1983	Ex Midland Red Coaches, 1986
1014	A656VDA	Leyland Tiger TRCTL11/3R	Plaxton Paramount 3500	C48FT	1983	Ex Midland Red Coaches, 1986
1015	A657VDA	Leyland Tiger TRCTL11/3R	Plaxton Paramount 3500	C48FT	1983	Ex Midland Red Coaches, 1986
1016	A658VDA	Leyland Tiger TRCTL11/3R	Plaxton Paramount 3200	C50FT	1983	Ex Midland Red Coaches, 1986
1017	B566BOK	Leyland Tiger TRCTL11/3RH	Duple Caribbean 2	C48FT	1984	Ex Midland Red Coaches, 1986
1018	B567BOK	Leyland Tiger TRCTL11/3RH	Duple Caribbean 2	C48FT	1984	Ex Midland Red Coaches, 1986
1019	B568BOK	Leyland Tiger TRCTL11/3RH	Duple Caribbean 2	C48FT	1984	Ex Midland Red Coaches, 1986
1020	C985HOX	Leyland Tiger TRCTL11/3RZ	Duple 340	C48FT	1986	Ex Midland Red Coaches, 1986
1021	C986HOX	Leyland Tiger TRCTL11/3RZ	Duple 340	C48FT	1986	Ex Midland Red Coaches, 1986
1022	C987HOX	Leyland Tiger TRCTL11/3RZ	Duple 340	C48FT	1986	Ex Midland Red Coaches, 1986

Thirteen Dennis Darts joined Midland Red West in 1995, and these carry Plaxton Pointer bodies fitted with high-back seating. The village of Cleobury Mortimer is the setting for this picture of 313, N313XAB as the bus heads for the south Shropshire town of Ludlow. *Richard Godfrey*

1101-1150		Leyland Lynx LX2R11C15Z4R		Leyland Lynx		B49F		1990	
1101	G101HNP	**1111**	G111HNP	**1121**	G121HNP	**1131**	G131HNP	**1141**	G141HNP
1102	G102HNP	**1112**	G112HNP	**1122**	G122HNP	**1132**	G132HNP	**1142**	G142HNP
1103	G103HNP	**1113**	G113HNP	**1123**	G123HNP	**1133**	G133HNP	**1143**	G143HNP
1104	G104HNP	**1114**	G114HNP	**1124**	G124HNP	**1134**	G134HNP	**1144**	G144HNP
1105	G105HNP	**1115**	G115HNP	**1125**	G125HNP	**1135**	G135HNP	**1145**	G145HNP
1106	G106HNP	**1116**	G116HNP	**1126**	G126HNP	**1136**	G136HNP	**1146**	G146HNP
1107	G107HNP	**1117**	G117HNP	**1127**	G127HNP	**1137**	G137HNP	**1147**	G147HNP
1108	G108HNP	**1118**	G118HNP	**1128**	G128HNP	**1138**	G138HNP	**1148**	G148HNP
1109	G109HNP	**1119**	G119HNP	**1129**	G129HNP	**1139**	G139HNP	**1149**	G149HNP
1110	G110HNP	**1120**	G120HNP	**1130**	G130HNP	**1140**	G140HNP	**1150**	G150HNP

1301-1319		Mercedes-Benz L608D		PMT Hanbridge		B20F*		1985	*1301/4-6 are DP20F
1301	C301PNP	**1307**	C307PNP	**1311**	C311PNP	**1314**	C314PNP	**1317**	C317PNP
1304	C304PNP	**1308**	C308PNP	**1312**	C312PNP	**1315**	C315PNP	**1318**	C318PNP
1305	C305PNP	**1309**	C309PNP	**1313**	C313PNP	**1316**	C316PNP	**1319**	C319PNP
1306	C306PNP	**1310**	C310PNP						

1320	C320PNP	Mercedes-Benz L608D	Alexander AM	B20F	1986

1321-1361		Mercedes-Benz L608D		Robin Hood		B20F		1985-86	
1321	C321PNP	**1330**	C330PNP	**1338**	C338PNP	**1346**	C346PNP	**1354**	C354PNP
1322	C322PNP	**1331**	C331PNP	**1339**	C339PNP	**1347**	C347PNP	**1355**	C355PNP
1323	C323PNP	**1332**	C332PNP	**1340**	C340PNP	**1348**	C348PNP	**1356**	C356PNP
1324	C324PNP	**1333**	C333PNP	**1341**	C341PNP	**1349**	C349PNP	**1357**	C357PNP
1325	C325PNP	**1334**	C334PNP	**1342**	C342PNP	**1350**	C350PNP	**1358**	C358PNP
1326	C326PNP	**1335**	C335PNP	**1343**	C343PNP	**1351**	C351PNP	**1359**	C359PNP
1327	C327PNP	**1336**	C336PNP	**1344**	C344PNP	**1352**	C352PNP	**1360**	C360PNP
1328	C328PNP	**1337**	C337PNP	**1345**	C345PNP	**1353**	C353PNP	**1361**	C361RUY
1329	C329PNP								

149 REDDITCH
MIDLAND RED WEST
1136
Leyland
G136 HNP

436 BRIDGNORTH
Midland Red West
SHROPSHIRE
BUS
SUPER
LOW FLOOR
BUS
341
Midland Red West
DENNIS
PLAXTON
N341 EUY

Two Leyland Nationals from sister company Western National remain in service with Midland Red West. Seen in full livery is 756, AFJ756T. *Robert Edworthy*

1362-1382		Mercedes-Benz L608D		Reeve Burgess		B20F	1986		
1362	C362RUY	**1366**	C366RUY	**1370**	C370RUY	**1373**	C373RUY	**1376**	C376RUY
1363	C363RUY	**1367**	C367RUY	**1371**	C371RUY	**1374**	C374RUY	**1381**	C381RUY
1364	C364RUY	**1368**	C368RUY	**1372**	C372RUY	**1375**	C375RUY	**1382**	C382RUY
1365	C365RUY	**1369**	C369RUY						

1385-1404		Mercedes-Benz L608D		Robin Hood		B20F	1986		
1385	C385RUY	**1389**	C389RUY	**1393**	C393RUY	**1397**	C397RUY	**1401**	C401RUY
1386	C386RUY	**1390**	C390RUY	**1394**	C394RUY	**1398**	C398RUY	**1402**	C402RUY
1387	C387RUY	**1391**	C391RUY	**1395**	C395RUY	**1399**	C399RUY	**1403**	C403RUY
1388	C388RUY	**1392**	C392RUY	**1396**	C396RUY	**1400**	C400RUY	**1404**	C404RUY

1406-1439		Mercedes-Benz 609D		Reeve Burgess		B20F	1987-88		
1406	E406HAB	**1413**	E413KUY	**1420**	E420KUY	**1427**	E427KUY	**1434**	E434KUY
1407	E407HAB	**1414**	E414KUY	**1421**	E421KUY	**1428**	E428KUY	**1435**	E435KUY
1408	E408HAB	**1415**	E415KUY	**1422**	E422KUY	**1429**	E429KUY	**1436**	E436KUY
1409	E409HAB	**1416**	E416KUY	**1423**	E423KUY	**1430**	E430KUY	**1437**	E437KUY
1410	E410HAB	**1417**	E417KUY	**1424**	E424KUY	**1431**	E431KUY	**1438**	E438KUY
1411	E411HAB	**1418**	E418KUY	**1425**	E425KUY	**1432**	E432KUY	**1439**	E439KUY
1412	E412KUY	**1419**	E419KUY	**1426**	E426KUY	**1433**	E433KUY		

1440	C475BHY	Mercedes-Benz L608D	Reeve Burgess	B20F	1986	Ex Bristol, 1988

Overleaf top: **In 1990 a batch of fifty Leyland Lynx were delivered to Midland Red West, most of which worked from the then Birmingham base. Now scattered to various depots, they currently retain their original liveries. Seen here is 1136, G136HNP.**
Overleaf bottom: **Midland Red West's latest vehicle has been acquired to meet the needs of a recently won Shropshire Bus contract. This calls for a low floor vehicle, and Dennis Dart 341, N341EUY is seen at Bridgnorth appropriately lettered for it's work. This vehicle is the only bus to date with the new style of lettering to FirstBus corporate style.**

The mainstay of the minibus fleet is the Mercedes-Benz L608D van conversion most of which have conversions by Reeve Burgess. Typical is 1446, C586SHC, photographed in Bromsgrove. One of nine acquired from Southdown in 1988, it is from a batch now spread far and wide with several operating in New Zealand with Stagecoach Wellington. *David Cole*

1441-1449 Mercedes-Benz L608D | Reeve Burgess | B20F | 1986 | Ex Southdown, 1988

1441	C581SHC	**1443**	C583SHC	**1445**	C585SHC	**1447**	C587SHC	**1449**	C589SHC
1442	C582SHC	**1444**	C584SHC	**1446**	C586SHC	**1448**	C588SHC		

1450	C788FRL	Mercedes-Benz L608D	Reeve Burgess	B20F	1986	Ex Western National, 1989
1451	C790FRL	Mercedes-Benz L608D	Reeve Burgess	B20F	1986	Ex Western National, 1989
1463	D763KWT	Mercedes-Benz 609D	Reeve Burgess	B20F	1987	Ex SWT, 1994

1476-1499 Mercedes-Benz L608D | Reeve Burgess | B20F | 1986 | Ex City Line, 1993

1476	C476BHY	**1487**	C487BHY	**1490**	C490BHY	**1492**	C492BHY	**1498**	C498BHY
1477	C477BHY	**1488**	C488BHY	**1491**	C491BHY	**1497**	C497BHY	**1499**	C499BHY
1483	C483BHY								

Previous Registrations:

Q276UOC	BVP804V	Q553UOC	LOA843X

Livery: Red and cream; red (Midland Red Coaches) 1001-7/13-22, 1301.

NEWBURY COACHES

KR, KM & AR Powell, Lower Road Trading Estate, Ledbury, Herefordshire, HR8 2DJ

MDG193W	Bedford YMT (Cummins)	Plaxton Supreme IV Express	C53F	1980	Ex Berline, Gloucester, 1984
C402XFO	Bedford YNV Venturer	Plaxton Paramount 3200 II	C57F	1986	
C355ALJ	Bedford YNV Venturer	Plaxton Paramount 3200 II	C57F	1986	Ex Brixham Coaches, 1989
E830EUT	Bedford YNV Venturer	Plaxton Paramount 3200 III	C57F	1988	Ex Wainfleet, Nuneaton, 1990
E833EUT	Bedford YNV Venturer	Plaxton Paramount 3200 III	C57F	1988	Ex Wainfleet, Nuneaton, 1991
E755HJF	Dennis Javelin 12SDA1907	Duple 320	C57F	1988	Ex Reliance, Benfleet, 1996
F791GNA	Leyland Tiger TRCTL11/3ARZ	Duple 320	C53F	1989	Ex Metroline, 1996
F803KCJ	Dennis Javelin 12SDA1907	Plaxton Paramount 3200 III	C52FT	1988	
G327SVV	Leyland Tiger TRCL10/3ARZM	Plaxton Paramount 3200 III	C53F	1990	Ex Wainfleet, Nuneaton, 1996
J663CVJ	Leyland Tiger TRCTL11/3AR	Plaxton Paramount 3200 III	C57F	1992	
DWK407T	Bedford YMT	Plaxton Supreme IV	C53F	1979	Ex Bryants Coaches, Williton, 1994
G518LWU	Volvo B10M-60	Plaxton Paramount 3500 III	C50F	1990	Ex Wallace Arnold, 1995
K830HVJ	Renault Master T35D	Pearl	M16	1993	
L424ANP	LDV400	LDV	M16	1994	
M737RCJ	Dennis Javelin 12SDA2125	Plaxton Premiere 320	C57F	1994	
M219TCJ	Mercedes-Benz 609D	Autobus Classique	B25F	1995	

Previous Registrations:
G327SVV G879BKV, MIW5786

Livery: White and blue

NICHOLLS

D M Nicholls, Little Newlands, Garway, Herefordshire, HR2 8RF

UJV831	Leyland Leopard PSU3E/4R	Plaxton Supreme IV (1979)	C49F	1965	Ex Crown Coaches, Bristol, 1988
MFV31T	Leyland Leopard PSU4E/4R	East Lancashire	B47F	1978	Ex Burnley & Pendle, 1996
FDV804V	Leyland Leopard PSU3F/4R	Plaxton Supreme IV Express	C49FT	1980	Ex Devon General, 1990

Previous Registrations:
UJV831 CHA101C, WFB510T

Livery: Various

Newbury Coaches operate from Ledbury in Herefordshire and employ a white and blue livery. The 1994 arrival is M737RCJ, a Dennis Javelin with Plaxton Première 320 bodywork seen here outside the depot. *Robert Edworthy*

Still showing the Pendle Witch is MFV31T in the Nicholls fleet which comprises of a trio of Leyland Leopards, the others being coaches with Plaxton Supreme bodies. The principal use of the East Lancashire-bodied Leopard is school duties from its base less than 2km from the Welsh Border. *Robert Edworthy*

THE OXFORD BUS COMPANY

City of Oxford Motor Services Ltd, 395 Cowley Road, Oxford, OX4 2DJ

A member of Go-Ahead Group Ltd.

Depots : Cowley Road, Oxford and Newlands Road, High Wycombe

50-55 Dennis Javelin 12SDA2118 | Plaxton Premiére 320 | C53F | 1992

50	K750UJO	**52**	K752UJO	**53**	K753UJO	**54**	K754UJO	**55**	K755UJO
51	K751UJO								

130-134 DAF MB230LT615 | Plaxton Paramount 3500 III | C53F | 1988

130	E130YUD	**131**	E131YUD	**132**	E132YUD	**133**	E133YUD	**134**	E134YUD

135-139 DAF SB3000DKV601 | Plaxton Paramount 3500 III | C53F | 1989

135	F135LJO	**136**	F136LJO	**137**	F137LJO	**138**	F138LJO	**139**	F139LJO

140	J140NJO	DAF SB2305DHS585	Plaxton Paramount 3200 III	C53F	1991
141	J141NJO	DAF SB2305DHS585	Plaxton Paramount 3200 III	C53F	1991

150-155 Volvo B10M-62 | Plaxton Premiére 350 | C53F | 1993

150	L150HUD	**152**	L152HUD	**153**	L153HUD	**154**	L154HUD	**155**	L155HUD
151	L151HUD								

156	N156BFC	Volvo B10M-62	Plaxton Premiere 350	C53F	1995
157	N157BFC	Volvo B10M-62	Plaxton Premiere 350	C53F	1995
158	N158BFC	Volvo B10M-62	Plaxton Premiere 350	C53F	1995

159-163 Volvo B10M-60 | Plaxton Paramount 3500 III | C51F | 1991 | Ex Shearings, 1995

159	UJI1759	**160**	UJI1760	**161**	UJI1761	**162**	UJI1762	**163**	UJI1763

201-224 Leyland Olympian ONLXB/1R* | Eastern Coach Works | H47/28D | 1982-83 | *221 is ONLXC/1R

201	VJO201X	**206**	VJO206X	**211**	WWL211X	**216**	BBW216Y	**221**	CUD221Y
202	VJO202X	**207**	WWL207X	**212**	WWL212X	**217**	BBW217Y	**222**	CUD222Y
203	VJO203X	**208**	WWL208X	**213**	BBW213Y	**218**	BBW218Y	**223**	CUD223Y
204	VJO204X	**209**	WWL209X	**214**	BBW214Y	**219**	CUD219Y	**224**	CUD224Y
205	VJO205X	**210**	WWL210X	**215**	BBW215Y	**220**	CUD220Y		

225-229 Leyland Olympian ONLXB/1RH | Alexander RL | H47/26D | 1988

225	E225CFC	**226**	E226CFC	**227**	E227CFC	**228**	E228CFC	**229**	E229CFC

The Oxford Bus Company is part of the Go-Ahead Group which includes Brighton & Hove and London Central as well as the fleets in the north east of England. While these are still operated separately, the purchasing policy of the parent company is starting to be noticed. *Opposite, top* **is 519, M519VJO, a Dennis Dart with Plaxton Verde bodywork delivered in 1995. The 1996 intake has moved to Volvo B10Bs with Plaxton Verde bodywork.** *M E Lyons*
A major part of the Oxford traffic programme is the encouragement of visitors and workers to use the Park & Ride facilities provided by the council. Ample notice to drivers is given when approaching the city and frequent services from the car parks into the centre are offered. Vehicles of the Oxford Bus Company that are used on the services carry special liveries. *Opposite, bottom* **is Leyland Titan 958, KYV370X, with bodywork built at Lillyhall alongside the National 2s.**

Marston
J.R. HOSPITAL 13
Cityline
519
Cityline
M519 VJO

Park&Ride
300 REDBRIDGE
Park&Ride
OXFORD BUS COMPANY
958
KYV 370X
OMFORT

Twenty-eight Volvo B10Bs with Plaxton Verde bodies are the latest single-deck buses with the Oxford Bus Company. Seen here is 612, N612FJO as it passes St Giles in Oxford. The Verde is expected to be replaced next year with a new model from the Henly Group (Plaxton's parent company) though this may be built at the Northern Counties bus assembly plant where production of large buses and double-decks is concentrated. *Malc McDonald*

230-235		Leyland Olympian ON2R50G16Z4	Alexander RL	H47/28F*	1990	*234/5 are H47/29F

230	G230VWL	**232**	G232VWL	**233**	G233VWL	**234**	G234VWL	**235**	G235VWL
231	G231VWL								

236	FWL778Y	Leyland Olympian ONLXB/1R	Eastern Coach Works	H45/32F	1983	Ex UKAEA, Harwell, 1991
237	FWL779Y	Leyland Olympian ONLXB/1R	Eastern Coach Works	H45/32F	1983	Ex UKAEA, Harwell, 1991
239	FWL781Y	Leyland Olympian ONLXB/1R	Eastern Coach Works	H45/32F	1983	Ex UKAEA, Harwell, 1991
240	D822UTF	Leyland Olympian ONLXB/1RH	Eastern Coach Works	CH39/21F	1986	Ex The Bee Line, 1990
241	D823UTF	Leyland Olympian ONLXB/1RH	Eastern Coach Works	CH39/21F	1986	Ex The Bee Line, 1990
242	D824UTF	Leyland Olympian ONLXB/1RH	Eastern Coach Works	CH39/21F	1986	Ex The Bee Line, 1990

301-305		Leyland Lynx LX112L10ZR1S	Leyland Lynx	B49F	1988	Ex The Bee Line, 1990

301	F556NJM	**302**	F557NJM	**303**	F558NJM	**304**	F559NJM	**305**	F560NJM

384	VPF296S	Leyland National 11351A/1R		B45F	1978	Ex The Bee Line, 1990
385	JWV127W	Leyland National 2 NL116L11/1R		B52F	1980	Ex Brighton & Hove, 1996
386	JWV128W	Leyland National 2 NL116L11/1R		B52F	1980	Ex Brighton & Hove, 1996

At one time the Wycombe Bus fleet was identified by the addition of 1000 to the Oxford numbers. All now share the same series as shown on former London Buses 377, THX177S, a Leyland National which moved from the capital to this South Midlands operation in 1993. As can be seen in the picture, high-back seating is fitted in this vehicle. *Malc McDonald*

443	SNJ592R	Bristol VRT/SL3/6LXB	Eastern Coach Works	H43/27D	1977	Ex Brighton & Hove, 1995
444	AAP651T	Bristol VRT/SL3/6LXB	Eastern Coach Works	H43/27D	1978	Ex Brighton & Hove, 1995
446	EAP989V	Bristol VRT/SL3/6LXB	Eastern Coach Works	H43/27D	1980	Ex Brighton & Hove, 1995
447	EAP999V	Bristol VRT/SL3/6LXB	Eastern Coach Works	H43/27D	1980	Ex Brighton & Hove, 1995
448	MRJ8W	Bristol VRT/SL3/6LXB	Eastern Coach Works	DPH41/29F	1980	Ex Mayne, Manchester, 1991
449	MRJ9W	Bristol VRT/SL3/6LXB	Eastern Coach Works	DPH41/29F	1980	Ex Mayne, Manchester, 1991

450-462

Bristol VRT/SL3/6LXB — Eastern Coach Works — H43/31F — 1976 — Ex The Bee Line, 1990

450	GGM110W	**453**	HJB453W	**456**	HJB456W	**458**	HJB458W	**461**	HJB461W
451	HJB451W	**454**	HJB454W	**457**	HJB457W	**459**	HJB459W	**462**	HJB462W
452	HJB452W	**455**	HJB455W						

501-520

Dennis Dart 9SDL3054 — Marshall C37 — B36D — 1995

501	M501VJO	**505**	M505VJO	**509**	M509VJO	**513**	M513VJO	**517**	M517VJO
502	M502VJO	**506**	M506VJO	**510**	M510VJO	**514**	M514VJO	**518**	M518VJO
503	M503VJO	**507**	M507VJO	**511**	M511VJO	**515**	M515VJO	**519**	M519VJO
504	M504VJO	**508**	M508VJO	**512**	M512VJO	**516**	M516VJO	**520**	M520VJO

521-527

Dennis Dart SLF — Plaxton Pointer — B36F* — 1995 — *B31F plus wheelchairs.

521	N521MJO	**523**	N523MJO	**525**	P525	**526**	P526	**527**	P527
522	N522MJO	**524**	N524MJO						

601-624

Volvo B10B-58 — Plaxton Verde — B51F — 1995-96

601	N601FJO	**606**	N606FJO	**611**	N611FJO	**616**	N616FJO	**621**	N621FJO
602	N602FJO	**607**	N607FJO	**612**	N612FJO	**617**	N617FJO	**622**	N622FJO
603	N603FJO	**608**	N608FJO	**613**	N613FJO	**618**	N618FJO	**623**	N623FJO
604	N604FJO	**609**	N609FJO	**614**	N614FJO	**619**	N619FJO	**624**	N624FJO
605	N605FJO	**610**	N610FJO	**615**	N615FJO	**620**	N620FJO		

625	N413NRG	Volvo B10B-58	Plaxton Verde	B51F	1995
626	N414NRG	Volvo B10B-58	Plaxton Verde	B51F	1995
627	N415NRG	Volvo B10B-58	Plaxton Verde	B51F	1995
628	N416NRG	Volvo B10B-58	Plaxton Verde	B51F	1995

Oxford suffers greatly from the effects of heavy traffic and much is being done to address the problem. A joint venture with Oxfordshire County Council and Southern Electric led to the operation of four electric MetroRiders on the City Circuit. These carry a livery of grey, silver and blue. Shown here is 803, L803HJO. *Phillip Stephenson*

During 1993, the Oxford Bus Company took into stock twenty-five Leyland Titans that replaced elderly Bristol VRs. Shown here in Oxford's High Street is 974, NUW635Y. The city has seen much competition between operators and substantial pedal-power from the students. *Richard Godfrey*

701	G621XLO	Mercedes-Benz 811D	Reeve Burgess Beaver	B29F	1989	Ex London Central, 1995
702	G222KWE	Mercedes-Benz 811D	Reeve Burgess Beaver	B26F	1989	Ex London Central, 1995
703	H189RWF	Mercedes-Benz 811D	Reeve Burgess Beaver	B29F	1990	Ex London Central, 1995
704	H191RWF	Mercedes-Benz 811D	Reeve Burgess Beaver	B29F	1990	Ex London Central, 1995

750-762 MCW MetroRider MF150/26* — MCW — B25F — 1987 — *750 is MF150/13, *758-62 are MF150/51

750	D750SJO	**753**	E753VJO	**756**	E756VJO	**759**	E759XWL	**761**	E761XWL
751	E751VJO	**754**	E754VJO	**757**	E757VJO	**760**	E760XWL	**762**	E762XWL
752	E752VJO	**755**	E755VJO	**758**	E758XWL				

763	F763LBW	MCW MetroRider MF150/114	MCW	B25F	1989

764-768 MCW MetroRider MF150/109* — MCW — B23F — 1989 — Ex Merthyr Tydfil, 1989; *768 is MF150/105

764	F501ANY	**765**	F502ANY	**766**	F503ANY	**767**	F504ANY	**768**	F505CBO

769-783 Optare MetroRider MR09 — Optare — B25F* — 1990 — *769-74 are B23F

769	G769WFC	**772**	G772WFC	**775**	G775WFC	**778**	G778WFC	**781**	G781WFC
770	G770WFC	**773**	G773WFC	**776**	G776WFC	**779**	G779WFC	**782**	G782WFC
771	G771WFC	**774**	G774WFC	**777**	G777WFC	**780**	G780WFC	**783**	G783WFC

801	L801HJO	Optare MetroRider MREL	Optare	B18F	1993	On loan from Southern Electric
802	L802HJO	Optare MetroRider MREL	Optare	B18F	1993	On loan from Southern Electric
803	L803HJO	Optare MetroRider MREL	Optare	B18F	1993	On loan from Southern Electric
804	L804HJO	Optare MetroRider MREL	Optare	B18F	1993	On loan from Southern Electric

950-975 Leyland Titan TNLXB2RRSp — Leyland — H44/26D* — 1981-83 — Ex London Buses, 1993; *955/62/3/70/74 are H44/24D

950	GYE280W	**955**	KYV452X	**960**	OHV711Y	**965**	OHV783Y	**970**	KYN308X
951	KYV516X	**956**	KYV519X	**961**	KYV524X	**966**	KYV381X	**971**	KYV457X
952	KYV300W	**957**	KYN291X	**962**	KYV530X	**967**	KYV392X	**973**	KYV493X
953	KYV317X	**958**	KYV370X	**963**	OHV727Y	**968**	NUW661Y	**974**	NUW635Y
954	KYV328X	**959**	NUW667Y	**964**	OHV745Y	**969**	KYV510X	**975**	A869SUL

999	PWL999W	Leyland Olympian B45/TL11/2R	Alexander RL	H50/34D	1980	Ex Leyland demonstrator, 1987

Previous Registrations:

UJI1759	H959DRJ	UJI1761	H957DRJ	UJI1763	H958DRJ
UJI1760	H954DRJ	UJI1762	H960DRJ		

Liveries: Red, white and blue (Buses); dark blue, yellow and white (City Link coaches); green, white and blue (Park & Ride)

Operating units:
Wycombe Bus Company: 230-242, 301-5/84-6, 443-62, 521-7, 701-4/69-82.
Oxford Bus Company: remainder

PEARCES

C & M Pearce, 39 Abingdon Road, Dorchester-on-Thames, Oxfordshire OX9 8JZ

Depot :Tower Road, Berinsfield

JUD597W	Ford R1014	Plaxton Supreme IV Express	C45F	1980	Ex House, Watlington, 1987
H634HBW	Toyota Hiace	Toyota	M8	1990	
J100OFC	Toyota Coaster HDB30R	Caetano Optimo II	C18F	1991	
K100OMP	Iveco 315-8-17	Lorraine	C30F	1992	
K200OMP	Iveco 315-8-17	Lorraine	C30F	1992	
L11VWL	Plaxton 425	Lorraine	C53F	1993	
M489HBC	Toyota Coaster HZB50R	Caetano Optimo III	C21F	1994	
M968RWL	Bova FLC12.280	Bova Futura Club	C53F	1995	
N108BHL	Mercedes-Benz 711D	Plaxton Beaver	C25F	1995	
N129MBW	Scania K113CRB	Van Hool Alizée	C F	1996	
N649KWL	Dennis Javelin 12SDA2159	Plaxton Premiere 350	C53F	1996	

Livery: White, yellow and red

Pearces fleet is of interest as it operates such varied capacities with a touch of the unusual. Certainly incoming tour work and private hire parties of all sizes are catered for. The small coaches employed, rather than large minibuses, give the impression of quality and comfort. Representing the fleet is J200OWP, a DAF SB2305 with Plaxton Paramount 3200 bodywork photographed when escorting a school party to London. *Phillip Stephenson*

PRIMROSE MOTORS

Primrose Motor Services (Leominster) Ltd, Worcester Road, Leominster, Herefordshire, HR6 8AR

OFR934T	Bedford YMT	Duple Dominant II	C53F	1979	Ex Stockcross Hire, 1988
BNO698T	Bedford YRT	Duple Dominant II Express	C53F	1979	Ex Yeomans, Hereford, 1995
AFO245V	Ford R1114	Duple Dominant II	C53F	1979	
LVS433V	Bedford YMT	Plaxton Supreme IV	C53F	1980	Ex Wivey Coaches, Bagborough, 1996
YLW897X	Bedford YMQ	Lex Maxeta	B37F	1981	Ex Yeomans, Hereford, 1995
MUV837X	Leyland Leopard PSU5C/4R	Duple Dominant IV	C53F	1982	Ex King Offa, Westbury, 1996
DHA986Y	Ford R1115	Plaxton Paramount 3200	C53F	1983	Ex Daisy, Broughton, 1991
A416DCN	Bedford YNT	Plaxton Paramount 3200	C53F	1984	Ex Chambers, Prees, 1990
B700SFO	Bedford VAS5	Plaxton Supreme IV	C29F	1984	Ex Yeomans, Hereford, 1996
XOI1908	Dennis Javelin 12SDA1907	Plaxton Paramount 3200 III	C53F	1988	Ex Evans, Tregaron, 1995
F833LCJ	Peugeot-Talbot Pullman	Talbot	B22F	1988	
F834LCJ	Peugeot-Talbot Pullman	Talbot	B22F	1988	
IIL1361	Volvo B10M-60	Plaxton Paramount 3500 III	C49FT	1989	Ex Happy Days, Woodseaves, 1996
G659TCJ	Dennis Javelin 12SDA1912	Plaxton Paramount 3200 III	C49F	1990	Ex Yeomans, Hereford, 1994

Previous Registrations:

A416DCN	A264BTY, UPP938	YLW897X	LCY301X, 43FJF, GGK237X, RIB7017
IIL1361	F362URF, 917DBO, Fxxxxxx	XOI1908	E136PLJ

Livery: Primrose and white.

The control of Primrose Motor Services has passed to Hereford-based Yeoman-Canyon Travel since the previous edition of this Bus Handbook. Consequently, many changes to the fleet have occurred. Now the oldest vehicle in the fleet, Bedford YRT OFR934T was operating service 292 to Leominster when photographed. *Martin Grosberg*

Since the 1920s Pulham's have operated services from the Gloucestershire villages of Naunton and Bouton-on-the-Water with a fleet of coaches. During the 1970s and early 1980s the vehicles were built to grant-door specification and mostly comprised products from Plaxton. Shown here are two different types both photographed at Banbury. Above is NFH200W, a Leyland Leopard with Supreme IV Express body style while below is XDG614, a Leyland Tiger with an Express variant of the Paramount 3200 body. *Richard Godfrey*

PULHAM'S

Pulham & Sons (Coaches) Ltd, Station Road Garage, Bourton-on-the-Water, GL54 2EN

DDD200T	Leyland Leopard PSU3E/4R	Plaxton Supreme IV Express	C53F	1979	
NAD600W	Leyland Leopard PSU3E/4R	Plaxton Supreme IV Express	C53F	1980	
NFH200W	Leyland Leopard PSU3E/4R	Plaxton Supreme IV Express	C49F	1980	
VDG700X	Leyland Leopard PSU3F/4R	Plaxton Supreme IV Express	C49F	1982	
FDF965	Leyland Tiger TRCTL11/3R	Plaxton Paramount 3200	C57F	1983	
VDF365	Leyland Tiger TRCTL11/3R	Plaxton Paramount 3200	C57F	1983	
VAD141	Volvo B10M-56	Plaxton Paramount 3200	C53F	1983	Ex Smith, Tring, 1995
WDF946	Leyland Tiger TRCTL11/2R	Plaxton Paramount 3200 E	C53F	1984	Ex Pilcher, Strood, 1986
XDG614	Leyland Tiger TRCTL11/3R	Plaxton Paramount 3200 IIE	C53F	1986	
ODF561	Volvo B9M	Plaxton Paramount 3200 II	C37F	1986	Ex Tellings Golden Miller, Byfleet, 1989
LDD488	Volvo B10M-61	Plaxton Paramount 3500 III	C53F	1988	
UDF936	Volvo B10M-61	Plaxton Paramount 3500 III	C53F	1989	
F634FNA	Ford Transit VE6	Made-to-Measure	M12	1989	Ex Willock, Macclesfield, 1993
G680YLP	Ford Transit VE6	Dormobile	M16L	1990	Ex LB Harrow, 1996
PDF567	Volvo B10M-60	Plaxton Paramount 3200 III	C53F	1991	Ex Supreme, Coventry, 1993
HDF661	Volvo B10M-60	Plaxton Paramount 3200 III	C53F	1991	Ex Supreme, Coventry, 1993
H345KDF	Volvo B10M-60	Plaxton Paramount 3200 III	C55F	1991	
J914MDG	Volvo B10M-60	Plaxton Paramount 3200 III	C57F	1991	
K929VDF	Toyota Coaster HDB30R	Caetano Optimo II	C21F	1993	
N680RDD	Volvo B10M-62	Van Hool Alizée	C49FT	1996	
P	Toyota Coaster HZB51R	Caetano Optimo III	C21F	1996	

Previous Registrations:

FDF965	DDD122Y	PDF567	H156HAC	VAD141	A22NRO
HDF661	H155HAC	UDF936	F401UAD	WDF946	A948JAY
LDD488	F150RFH	VDF365	DAD600Y	XDG614	C71XDG
ODF561	C193CYO				

Livery: Cream and red.

Pulham's purchased a Toyota Coaster in 1993 and have a further example on order. These carry Caetano Optimo bodywork built in Portugal. Shown at rest in Cheltenham is K929VDF. The success in acquiring several Heritage index marks originating in Gloucestershire is noteable.
Richard Eversden

REDLINE

Redline Bus Co Ltd, 6 School Lane, Lickey End, Bromsgrove, Worcestershire, B60 1JE

111	D105TFT	Freight Rover Sherpa	Carlyle Citybus	B20F	1986	Ex Busways, 1990
112	G224EOA	Freight Rover Sherpa	Carlyle Citybus 2	B20F	1989	Ex Carlyle Bus, 1992
113	L113YAB	Iveco TurboDaily 59.12	ECC	B29F	1994	
114	N114YAB	Iveco TurboDaily 59.12	ECC	B29F	1995	
115	G865BPD	Iveco Daily 49.10	Carlyle Dailybus 2	B25F	1989	Ex London & Country (Horsham Buses), 1996

Livery: Red & White

ROGERS of MARTLEY

DG & MA Rogers, Rogonda, The Noak, Martley, Worcestershire, WR6 6PD

Depot : The Smithy, Martley.

MEF154J	Ford R226	Plaxton Elite II	C49F	1971	Ex Beeline, Hartlepool, 1997
MCJ500P	Bedford YLQ	Plaxton Supreme III	C45F	1976	Ex Youngs, Kempley, 1985
VPT965R	Bedford YMT	Duple Dominant	C53F	1976	Ex Parker, Hindolveston, 1991
50ABK	Bedford YMT	Plaxton Supreme III	C53F	1978	Ex Canyon, Hereford, 1983
SDR595	Bedford YMT	Duple Dominant II	C53F	1980	Ex Whittle, Highley, 1982
C975HOX	MCW Metroliner DR130/14	MCW	CH55/17DT	1986	Ex Midland Red West, 1991
C976HOX	MCW Metroliner DR130/14	MCW	CH55/17DT	1986	Ex Midland Red West, 1991
C977HOX	MCW Metroliner DR130/14	MCW	CH55/17DT	1986	Ex Midland Red West, 1991
GIL2782	Scania K112CRS	Plaxton Paramount 3200 II	C55F	1986	Ex Humming Bird, Bromley, 1995
E62EVJ	Bedford CFL	Steerdrive Parflo	M14	1987	
F830GKO	Scania K92CRB	Duple 320	C55F	1989	Ex Warren, Ticehurst, 1993
K196SFH	Toyota Coaster HDB30R	Caetano Optimo II	C21F	1992	Ex Davis Coaches, Minchinhampton, 1996

Previous Registrations:

50ABK	WCJ600T	GIL2782	D169VVO	SDR595	FUJ923V

Livery: Yellow, black and red.

N114YAB of Redline was loading passengers in Bromsgrove when photographed shortly after delivery. Redline operates two of these Iveco minibuses which carry bodywork by ECC, one of several re-incarnations of the Robin Hood coachbuilders. *David Cole*

When Scania dealer Stuart Johnson had 30 K92/K93 chassis bodied by Duple, the 320 only was chosen as it was envisaged the combination would fulfil a market need for a standard, high-capacity coach at a competitive price. Duple ceased production in 1990 though the 320 body style continued for a short period as a Plaxton 321 and built in Scarborough. *Robert Edworthy*

SARGEANTS

Sargeant Brothers Ltd, The Nook, Mill Street, Kington, Herefordshire, HR5 3AL

SGF483L	Bristol RELH6L	Plaxton Elite II	C51F	1974	Ex Davies Bros, Pencader, 1988
TBD172N	Bedford YRQ	Willowbrook 001	B45F	1974	Ex United Counties, 1982
KEY212P	Bedford SB5	Willowbrook	B42F	1976	Ex Gypsy Queen, Langley Park, 1991
SJO871T	Bedford YMT (Leyland)	Plaxton Supreme IV Express	C53F	1978	Ex Evans, Tregaron, 1992
AUJ732T	Bedford YMT	Duple Dominant II	C53F	1978	Ex Whittle, Highley, 1982
PUB13W	Bedford YLQ (Leyland)	Plaxton Supreme IV	C45F	1980	Ex Evans, Welshpool, 1992
LMS157W	Leyland Fleetline FE30AGR	Alexander AD	H44/31F	1980	Ex Clydeside, 1995
CIB9321	Bedford YMQ	Plaxton Supreme IV Express	C35F	1980	Ex Munro, Jedburgh, 1991
CIB7615	Bedford YNT	Plaxton Paramount 3200	C53F	1983	Ex Bedford demonstrator, 1988
C900JGA	Bedford YNT	Plaxton Paramount 3200 II	C53F	1986	Ex Evans, Tregaron, 1996
D829KWT	Freight Rover Sherpa	Dormobile	B16F	1987	Ex P&O Lloyd, Bagillt, 1990
C724DHW	Renault-Dodge S46	Reeve Burgess	B12FL	1986	Ex County of Avon, 1996
E95RWR	Mercedes-Benz 811D	Optare StarRider	B33F	1987	Ex Northern Bus, Anston, 1989
E168TWO	Freight Rover Sherpa	Carlyle Citybus 2	B20F	1988	Ex Warner, Tewkesbury, 1993
SIB6441	LAG G355Z	LAG Panoramic	C49FT	1988	Ex Shorey, Flitwick, 1995
F660EDH	Ford Transit VE6	Ford	M11	1988	Ex Self Drive Hire, 1991
L91WBX	Renault Trafic	Cymric	M16	1993	
KUX211W	Bedford YNT	Duple Dominant IV	C53F	1981	Ex Carnell, Sutton Bridge, 1995
HHU838X	Bedford YNT	Duple Dominant IV	C53F	1981	Ex Crickhowell Coaches, 1995
F324BRN	Ford Transit VE6	Ford	M14	1988	Ex Kellor, 1995
H753ELP	Ford Transit VE6	Ford	M11	1990	Ex L W Tours, Finchley, 1992
H918SCX	Peugeot-Talbot Pullman	Talbot	B F	1991	Ex private owner, 1995
M960VWY	Mercedes-Benz 0405	Optare Cityranger	B49F	1995	

Previous Registrations:

CIB7615	KVS655Y	CIB9321	XSH397V	SGF483L	40WMN
CIB7866	ANA109Y	CIB9344	NYG804M	SIB6141	E672NNV

Livery: Red and gold

SMITHS

Smith's Motors (Ledbury) Ltd, Coach Garage, Homend, Ledbury, Herefordshire, HR8 1BA

AHW8V	Leyland Leopard PSU3E/4R	Plaxton Supreme IV	C53F	1980	Ex Blue Iris, Nailsea, 1995
OKV399W	Bedford YMT	Plaxton Supreme IV Express	C53F	1980	Ex Evans, Tregaron, 1991
B470XBW	Fiat 60-10	Caetano Beja	C18F	1985	Ex Rainbow, Westbury, 1994
GIL1481	Leyland Royal Tiger RT	Plaxton Paramount 3500	C52F	1985	Ex Worthing Coaches, 1990
D124HMT	Leyland Royal Tiger RT	Van Hool Alizée	C53F	1987	Ex Greenslades, Exeter, 1993
D700BJF	Bedford YNV Venturer	Duple 320	C57F	1987	Ex Alpha, Brighton, 1988
D83WWV	Bedford YNV Venturer	Duple 320	C57F	1987	Ex Alpha, Brighton, 1989
E405MPX	Mercedes-Benz L307D	Devon Conversions	M12	1988	Ex DJS Rental, Fareham, 1994
E988NMK	Leyland Tiger TRCL10/3RZM	Duple 340	C55F	1988	Ex Alpha, Brighton, 1995
J733USF	Mercedes-Benz 811D	PMT Ami	C33F	1991	Ex Fiesta, Hayle, 1994

Livery: White, red, green and grey.

M960VWY arrived with Sargeants in 1995 in an attractive red and gold livery. This Mercedes-Benz 0405 is one of a growing number of full-size Mercedes-Benz buses to be brought into the country, though supplied in chassis form for British coachbuilders to body. This example carries the Optare Cityranger body that uses parts from the Delta with a standard Mercedes-Benz 0405 front.

Smiths of Ledbury operates services including one between their home town and Hereford. Pictured performing the duty is AHW8V a Leyland Leopard PSU3 with Plaxton Supreme IV coach bodywork. *Robert Edworthy*

SOUDLEY VALLEY

Soudley Valley Coaches Ltd, Soudley, Cinderford, GL14 2TX

ODM777L	Bedford YRT	Plaxton Elite III	C53F	1973	Ex Hollis Coaches, Queensferry, 1976
GBB997N	Leyland Leopard PSU4C/4R	Duple Dominant	C45F	1975	Ex Evans, Tregaron, 1990
GBB999N	Leyland Leopard PSU4C/4R	Duple Dominant	C45F	1975	Ex Tyne & Wear, 1981
OAD200P	Leyland Leopard PSU3C/4R	Plaxton Supreme III Express	C51F	1976	Ex Cottrells, Mitcheldean, 1989
SDD133R	Leyland Leopard PSU3E/4R	Plaxton Supreme III	C53F	1977	Ex Cottrells, Mitcheldean, 1986
NDF857W	Leyland Leopard PSU3F/5R	Duple Dominant II Express	C53F	1980	
NDF858W	Leyland Leopard PSU3F/5R	Duple Dominant II Express	C53F	1980	
VRY610X	Leyland Leopard PSU3E/4R	Duple Dominant IV	C53F	1982	Ex Baildon, Guiseley, 1988
D810PUK	Freight Rover Sherpa	Carlyle	B18F	1987	Ex Evans, Tregaron, 1992

Livery: Two tone grey and red.

SPRINGS TOURS

WR & GR Spring, 50 Lime Street, Evesham, Worcestershire, WR11 5AH

Depot: Hinton Station, Hinton-on-the-Green,

DRB60T	Leyland Leopard PSU5C/4R	Plaxton Supreme IV	C57F	1979	Ex Torr, Gedling, 1993
WJM815T	Leyland Leopard PSU3E/4R	Plaxton Supreme IV Express	C49F	1979	Ex Oakley Coaches, 1995
JNJ24V	Leyland Leopard PSU3E/4R	Plaxton Supreme IV Express	C48F	1980	Ex Andy James, Easton Grey, 1993
JDE973X	Leyland Tiger TRCTL11/3R	Plaxton Supreme VI Express	C57F	1982	Ex Llew Jones, Llanrwst, 1996
UAM207	Volvo B10M-61	Duple Laser	C57F	1983	Ex Harrod, Wormegay, 1989
A460CRM	Mercedes-Benz L608D	Reeve Burgess	C21F	1984	Ex Waters, Addlestone, 1992
9896EL	Volvo B10M-61	Plaxton Paramount 3500	C49FT	1984	
C516DND	Volvo B10M-61	Plaxton Paramount 3200 II	C53F	1986	Ex Wilson, Carnwath, 1994
C125DWR	Volvo B10M-61	Plaxton Paramount 3500 II	C49FT	1986	Ex Wallace Arnold, 1993
D978PJA	Renault-Dodge S56	Northern Counties	B20F	1987	Ex GMN, 1996
E833LNP	Volvo B10M-61	Caetano Algarve	C51FT	1988	
F868TNH	Volvo B10M-61	Caetano Algarve	C53F	1988	Ex Scancoaches, North Acton, 1995

Previous Registrations:

9896EL	A852AUY	JNJ24V	GWV927V, 413DCD	UAM207	MSU583Y

Livery: Cream and red.

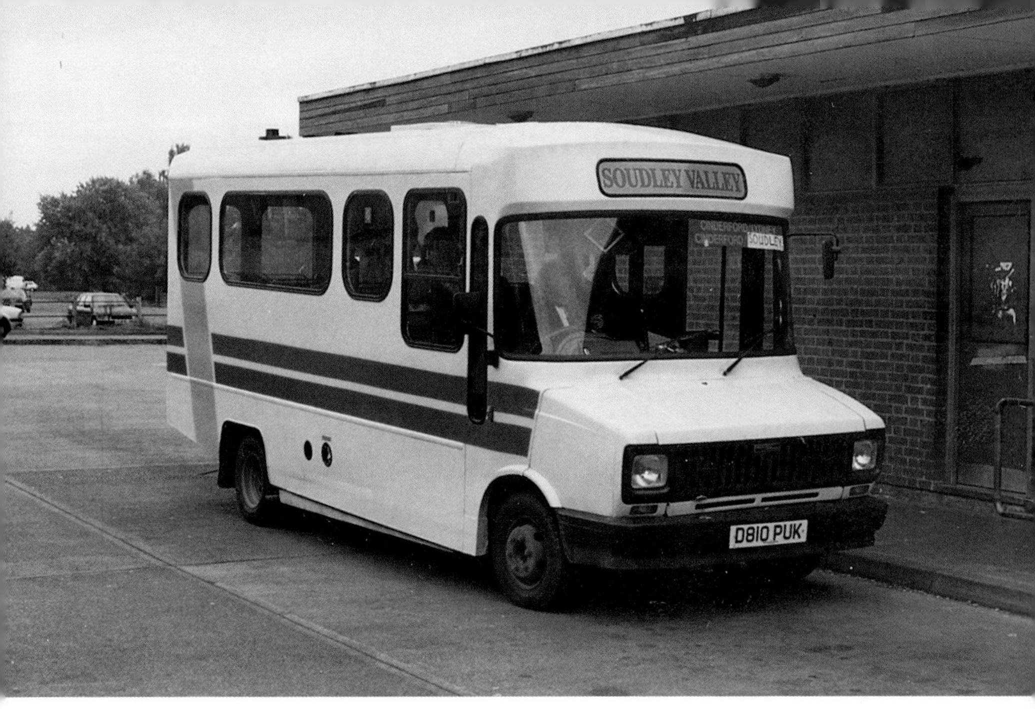

Sudley Valley is one of the longest established operators in the Forest of Dean having commenced in 1928. Inevitable changes in local industry have seen fleet strength reduced from 25 vehicles in the 1950s. A service from Cinderford to Lydney through Blakeney plus another to Gloucester is maintained. Freight Rover Sherpa D810PUK which carries a Carlyle bodywork was on the former route when photographed . *Robert Edworthy*

Straying far from its normal minibus duties is A460CRM of Springs Tours, photographed near Nelson's Column in London. The Mercedes-Benz L608D with Reeve Burgess conversion was joined in 1996 by a Renault-Dodge S56 minibus from Greater Manchester Buses North. *Colin Lloyd*

STAGECOACH MIDLAND RED

Midland Red (South) Ltd, Railway Terrace, Rugby, Warwickshire, CV21 3HS

Depots : Canal Street, Banbury; Rowley Drive, Coventry; Station Approach, Leamington Spa; Newtown Road, Nuneaton; Railway Terrace, Rugby and Avenue Farm, Stratford-on-Avon.

(Part of Stagecoach Holdings plc)

1	A75NAC	Leyland Tiger TRCTL11/2R	Plaxton Paramount 3200 E	C47FT	1983	
4	230HUE	Leyland Leopard PSU3E/4R	Plaxton Supreme IV Express	C49F	1980	Ex Midland Red North, 1981
5	331HWD	Leyland Leopard PSU3E/4R	Plaxton Supreme IV Express	C49F	1980	Ex Midland Red North, 1981
6	3273AC	Leyland Leopard PSU3E/4R	Plaxton Supreme IV Express	C46FT	1980	Ex Midland Red North, 1981
7	4012VC	Leyland Leopard PSU3E/4R	Plaxton Supreme IV Express	C49F	1979	Ex Premier Travel, 1991
15	NPA230W	Leyland Leopard PSU3E/4R	Plaxton Supreme IV Express	C53F	1981	Ex East Midland, 1994
16	YBO16T	Leyland Leopard PSU3E/2R	East Lancashire	B51F	1979	Ex G & G, Leamington, 1993
18	YBO18T	Leyland Leopard PSU3E/2R	East Lancashire	B51F	1979	Ex G & G, Leamington, 1993
19	A848VML	Leyland Leopard PSU3E/4R	Duple Dominant IV (1983)	C53F	1979	Ex Grey-Green, 1987
28	NAK28X	Leyland Leopard PSU3F/4R	Duple Dominant IV	C47F	1981	Ex East Midland, 1994
29	NAK29X	Leyland Leopard PSU3F/4R	Duple Dominant IV	C47F	1981	Ex East Midland, 1994
57	ANA435Y	DAF MB200DKTL600	Plaxton Paramount 3200	C51F	1983	Ex David R Grasby, 1995
58	9984PG	DAF MB200DKTL600	Duple Laser	C53F	1985	Ex Grey-Green, 1988
59	A6GGT	DAF SB2305DHTD585	Duple 320	C53F	1988	Ex Gray, Hoyland Common, 1990

60-65 Volvo B10M-60 Plaxton Paramount 3500 III C48FT 1990 Ex Wallace Arnold, 1993

60	G528LWU	**62**	G530LWU	**63**	G531LWU	**64**	G532LWU	**65**	G535LWU
61	G529LWU								

Two former East Midland Leyland Leopards are used by Stagecoach Midland Red on services from Banbury where both 29, NAK29X and sister vehicle 28 are based. *Malc McDonald*

While the re-paint programme continues there are a few vehicles still in the former scheme. Seen in Banbury is 6, 3273AC, a Leyland Leopard with Plaxton Supreme bodywork. *Richard Godfrey*

66	3063VC	Volvo B10M-60	Plaxton Paramount 3500 III	C49FT	1990	Ex Wallace Arnold, 1993
67	9258VC	Volvo B10M-60	Plaxton Paramount 3500 III	C49FT	1990	Ex Wallace Arnold, 1993
68	WSU293	Volvo B10M-60	Plaxton Paramount 3200 III	C49FT	1990	Ex Cheltenham & Gloucester, 1993
69	E315NWK	Volvo B10M-61	Ikarus Blue Danube	C53F	1987	Ex David R Gasby, 1995
74	4828VC	Leyland Tiger TRCTL11/3RH	Plaxton Paramount 3500 II	C51F	1985	Ex Sovereign, 1990
75	9737VC	Leyland Tiger TRCTL11/3R	Plaxton Paramount 3500 II	C51F	1985	Ex Sovereign, 1990
76	6253VC	Leyland Tiger TRCTL11/3RH	Plaxton Paramount 3200 II	C51F	1986	Ex Thames Transit, 1991
87	498FYB	Leyland Tiger TRCTL11/3R	Plaxton Paramount 3200	C50F	1983	Ex Cheltenham & Gloucester, 1993
88	A8GGT	Leyland Tiger TRCTL11/3R	Plaxton Paramount 3200 E	C57F	1983	Ex Cheltenham & Gloucester, 1993
89	A7GGT	Leyland Tiger TRCTL11/3RH	Plaxton Paramount 3200	C51F	1984	
90	552OHU	Leyland Tiger TRCTL11/3R	Plaxton Paramount 3200 E	C57F	1983	Ex Cheltenham & Gloucester, 1990
91	420GAC	Leyland Tiger TRCTL11/3R	Plaxton Paramount 3200 E	C46FT	1983	Ex Cheltenham & Gloucester, 1991

101-105 Dennis Dart SLF — Alexander Dash — B40F — 1996

101	P101HNH	**102**	P102HNH	**103**	P103HNH	**104**	P104HNH	**105**	P105HNH

201-216 Volvo B10M-55 — Alexander PS — DP48F* — 1995 — 206-212 are B49F

201	M201LHP	**205**	M205LHP	**208**	N208TDU	**211**	N211TDU	**214**	N214TDU
202	M202LHP	**206**	N206TDU	**209**	M209LHP	**212**	N212TDU	**215**	N215TDU
203	M203LHP	**207**	N207TDU	**210**	M210LHP	**213**	N213TDU	**216**	N216TDU
204	M204LHP								

300	E433YHL	Mercedes-Benz 709D	Reeve Burgess Beaver	B25F	1988	Ex Loftys, Bridge Trafford, 1993
301	G301WHP	Mercedes-Benz 709D	PMT	B25F	1989	
302	G302WHP	Mercedes-Benz 709D	PMT	B25F	1989	
303	G303WHP	Mercedes-Benz 709D	PMT	B25F	1989	
304	J304THP	Mercedes-Benz 709D	Alexander AM	B25F	1992	
305	J305THP	Mercedes-Benz 709D	Alexander AM	B25F	1992	
306	K306ARW	Mercedes-Benz 709D	Wright	B25F	1992	
307	L307SKV	Mercedes-Benz 709D	Wright	B25F	1993	

Prior to becoming part of Stagecoach, Midland Red South took delivery of a batch of Wright-bodied Mercedes-Benz 811s, several of which carried Stratford Blue names. Now in corporate livery, though still based at Stratford, is 409, J409PRW. This batch has several vehicles with hybrid seats that have a back that is higher than normal but not as high as used in a coach configuration. *Philip Lamb*

308-330		Mercedes-Benz 709D		Alexander Sprint		B23F	1994		
308	L308YDU	**313**	L313YDU	**318**	L318YDU	**323**	L323YDU	**327**	L327YKV
309	L309YDU	**314**	L314YDU	**319**	L319YDU	**324**	L324YDU	**328**	L328YKV
310	L310YDU	**315**	L315YDU	**320**	L310YDU	**325**	L325YDU	**329**	L329YKV
311	L311YDU	**316**	L316YDU	**321**	L321YDU	**326**	L326YKV	**330**	L330YKV
312	L312YDU	**317**	L317YDU	**322**	L322YDU				

331-346		Mercedes-Benz 709D		Alexander Sprint		B23F	1995		
331	M331LHP	**335**	M335LHP	**338**	M338LHP	**341**	M341LHP	**344**	M344LHP
332	M332LHP	**336**	M336LHP	**339**	M339LHP	**342**	M342LHP	**345**	M345LHP
334	M334LHP	**337**	M337LHP	**340**	M340LHP	**343**	M343LHP	**346**	M346LHP

347-372		Mercedes-Benz 709D		Alexander Sprint		B25F	1996		
347	N347AVV	**353**	N353AVV	**358**	N358AVV	**363**	N363AVV	**368**	N368AVV
348	N348AVV	**354**	N354AVV	**359**	N359AVV	**364**	N364AVV	**369**	N369AVV
349	N349AVV	**355**	N355AVV	**360**	N360AVV	**365**	N365AVV	**370**	N370AVV
350	N350AVV	**356**	N356AVV	**361**	N361AVV	**366**	N366AVV	**371**	N371AVV
351	N351AVV	**357**	N357AVV	**362**	N362AVV	**367**	N367AVV	**372**	N372AVV
352	N352AVV								

390	C705FKE	Ford Transit 190D	Dormobile	B16F	1986	Ex Stagecoach South, 1994
400	F71LAL	Mercedes-Benz 811D	Alexander AM	DP33F	1988	Ex Skills, Nottingham, 1991

Opposite top: **The number of Optare StarRiders in the Stagecoach group remains static mostly because the chassis type is compatible with the current minibus choice, the Mercedes-Benz. Shown here is 426, CSV219, a coach version acquired second-hand.**
Opposite bottom: **At the start of 1996, David R. Grasby coaches were identifiable as their fleet numbers were in the 3000 range. Now this operation has been included in the main business, 3435, ANA435Y, shown here, has been re-numbered 57 and received corporate livery within the main fleet.**

512
Stagecoach
426
CSV 219

DAVID R. GRASBY
3435
ANA 435Y

The 1994 intake for Stagecoach Midland Red included six Volvo B6s. These are currently shared between Leamington and Stratford where 452 was photographed at the end of its journey.
Philip Lamb

401-418	Mercedes-Benz 811D	Wright	B33F*	1991	*402/4/7-12 are DP33F *401/3/5/6/13/7/8 are B31F

401	H401MRW	**405**	H495MRW	**409**	J409PRW	**413**	J413PRW	**416**	J416PRW
402	H402MRW	**406**	H406MRW	**410**	J410PRW	**414**	J414PRW	**417**	J417PRW
403	H403MRW	**407**	J407PRW	**411**	J411PRW	**415**	J415PRW	**418**	J418PRW
404	H404MRW	**408**	J408PRW	**412**	J412PRW				

419	G115OGA	Mercedes-Benz 811D	Alexander AM	DP33F	1988	Ex Beaton, Blantyre, 1992

420-425	Mercedes-Benz 811D	Wright	B31F	1993

420	K420ARW	**422**	K422ARW	**423**	K423ARW	**424**	K424ARW	**425**	K425ARW
421	K421ARW								

426	CSV219	Mercedes-Benz 811D	Optare StarRider	C29F	1989	Ex Brents Coaches, Watford, 1993
427	H912XGA	Mercedes-Benz 814D	Reeve Burgess Beaver	DP31F	1990	Ex Loftys, Bridge Trafford, 1993

451-456	Volvo B6-9.9M	Alexander Dash	B40F	1994

451	L451YAC	**453**	L453YAC	**454**	L454YAC	**455**	L455YAC	**456**	L456YAC
452	L452YAC								

502	JOX502P	Leyland National 11351A/1R	B49F	1976	Ex Midland Red, 1981
503	JOX503P	Leyland National 11351A/1R	B49F	1976	Ex Midland Red, 1981
504	JOX504P	Leyland National 11351A/1R	B49DL	1976	Ex Midland Red, 1981
506	WAS765V	Leyland National 2 NL116l11/1R	B52F	1980	Ex Red & White, 1996
553	NOE553R	Leyland National 11351A/1R	B49F	1977	Ex Midland Red, 1981
554	NOE554R	Leyland National 11351A/1R	B49F	1977	Ex Cheltenham & Gloucester, 1994
571	NOE571R	Leyland National 11351A/1R	B49F	1977	Ex Midland Red, 1981
577	NOE577R	Leyland National 11351A/1R	B49F	1977	Ex Midland Red, 1981
578	NOE578R	Leyland National 11351A/1R	B49F	1977	Ex Midland Red, 1981

The single-deck fleet inherited by Stagecoach included many former G&G and Vanguard vehicles, most of which have now been sold. In their place the Leyland National fleet was increased with the arrival of units from Cheltenham & Gloucester. Photographed in Banbury is 600, SAE753S, dating from 1978. *Malc McDonald*

581	NOE581R	Leyland National 11351A/1R	B49F	1977	Ex Midland Red, 1981
582	NOE582R	Leyland National 11351A/1R (DAF)	B49F	1977	Ex Midland Red, 1981
586	NOE586R	Leyland National 11351A/1R	B49F	1977	Ex Midland Red, 1981
587	NOE587R	Leyland National 11351A/1R	B49F	1977	Ex Cheltenham & Gloucester, 1994
589	NOE589R	Leyland National 11351A/1R	B49F	1977	Ex Midland Red, 1981
590	NOE590R	Leyland National 11351A/1R	B49DL	1977	Ex Midland Red, 1981
591	YEU446V	Leyland National 10351B/1R	B44F	1981	Ex Cheltenham & Gloucester, 1994
592	NOE551R	Leyland National 11351A/1R	B49F	1976	Ex Midland Red, 1981
593	KHT122P	Leyland National 11351/1R	B52F	1976	Ex Cheltenham & Gloucester, 1994
594	VAE502T	Leyland National 10351B/1R	B44F	1979	Ex Cheltenham & Gloucester, 1994
595	GOL426N	Leyland National 11351/1R	B49F	1975	Ex Cheltenham & Gloucester, 1994
597	HEU122N	Leyland National 11351/1R	B52F	1975	Ex Cheltenham & Gloucester, 1994
598	KHT124P	Leyland National 11351/1R	B52F	1976	Ex Cheltenham & Gloucester, 1994
600	SAE753S	Leyland National 11351A/1R	B52F	1978	Ex Cheltenham & Gloucester, 1994

602-772 Leyland National 11351A/1R(DAF) B49F* 1977-80 Ex Midland Red, 1981
*624, 708 are B52F; 755/6 have LPG engines

602	NOE602R	**621**	PUK621R	**626**	PUK626R	**708**	TOF708S	**755**	XOV755T
603	NOE603R	**622**	PUK622R	**627**	PUK627R	**709**	TOF709S	**756**	XOV756T
604	NOE604R	**623**	PUK623R	**628**	PUK628R	**710**	TOF10S	**760**	XOV760T
605	NOE605R	**624**	PUK624R	**629**	PUK629R	**753**	XOV753T	**771**	BVP771V
606	NOE606R	**625**	PUK625R	**707**	TOF707S	**754**	XOV754T	**772**	BVP772V

802	SHH392X	Leyland National 2 NL116AL11/1R	B52F	1982	Ex Cheltenham & Gloucester, 1995
803	TAE639S	Leyland National 11351A/1R(DAF)	B52F	1978	Ex Cheltenham & Gloucester, 1995
808	BVP808V	Leyland National 2 NL116L11/1R	B49F	1980	Ex North Western, 1991
809	SVV589W	Leyland National 2 NL116L11/1R	B49F	1980	Ex Luton & District, 1991
816	BVP816V	Leyland National 2 NL116L11/1R (DAF)	B49F	1980	Ex Midland Red, 1981
817	BVP817V	Leyland National 2 NL116L11/1R (DAF)	B49F	1980	Ex Midland Red, 1981
818	BVP818V	Leyland National 2 NL116L11/1R (DAF)	B49F	1980	Ex Midland Red, 1981

Passing through Nuneaton heading for Coventry is Volvo B10M number 210, M210LHP. One of sixteen operated by Stagecoach Midland Red it carries an Alexander PS-type body. *Michael Fowler*

819	F661PWK	Leyland Lynx LX112L10ZR1R	Leyland	B51F	1988	
820	F660PWK	Leyland Lynx LX112L10ZR1R	Leyland	B51F	1988	

821-830 Iveco Daily 49.10 — Marshall C29 — B23F — 1993 — Ex Selkent, 1995

821	K521EFL	**823**	K523EFL	**825**	K525EFL	**827**	K527EFL	**829**	K529EFL
822	K522EFL	**824**	K524EFL	**826**	K526EFL	**828**	K528EFL	**830**	K530EFL

832	N182CMJ	Iveco Daily 59.12	Alexander	B29F	1995	Ex Iveco demonstrator, 1996
833	N183CMJ	Iveco Daily 59.12	Alexander	B29F	1996	Ex Iveco demonstrator, 1996

902-912 Leyland Olympian ONLXB/1R — Eastern Coach Works — H45/32F — 1983-84

902	A542HAC	**904**	A544HAC	**906**	A546HAC	**910**	B910ODU	**912**	B912ODU
903	A543HAC	**905**	A545HAC	**907**	A547HAC	**911**	B911ODU		

926	OBD842P	Bristol VRT/SL3/6LX	Eastern Coach Works	H43/31F	1976	Ex Circle Line, 1996
927	NHU671R	Bristol VRT/SL3/6LXB	Eastern Coach Works	H43/27D	1979	Ex Cheltenham & Gloucester, 1994
928	LHT725P	Bristol VRT/SL3/501(6LXB)	Eastern Coach Works	H39/31F	1976	Ex Cheltenham & Gloucester, 1994
929	NHU672R	Bristol VRT/SL3/6LXB	Eastern Coach Works	H43/27D	1979	Ex Cheltenham & Gloucester, 1994
930	LHT724P	Bristol VRT/SL3/501(6LXB)	Eastern Coach Works	H43/31F	1976	Ex Swindon & District, 1992
931	MAU145P	Bristol VRT/SL3/6LXB	Eastern Coach Works	H43/31F	1976	Ex Bluebird, 1993
932	CBV16S	Bristol VRT/SL3/501(6LXB)	Eastern Coach Works	H43/31F	1977	Ex Ribble, 1994
933	PEU516R	Bristol VRT/SL3/6LXB	Eastern Coach Works	H43/31F	1977	Ex Swindon & District, 1992
936	ONH846P	Bristol VRT/SL3/6LXB	Eastern Coach Works	H43/31F	1976	Ex Bluebird, 1993
937	DWF195V	Bristol VRT/SL3/6LXB	Eastern Coach Works	H43/31F	1979	Ex East Midland, 1994
939	DWF194V	Bristol VRT/SL3/6LXB	Eastern Coach Works	H43/31F	1979	Ex East Midland, 1994
940	PEU511R	Bristol VRT/SL3/6LXB	Eastern Coach Works	DPH43/31F	1977	Ex Badgerline, 1993
941	GTX746W	Bristol VRT/SL3/501	Eastern Coach Works	H43/31F	1980	Ex Red & White, 1993
943	GTX754W	Bristol VRT/SL3/501	Eastern Coach Works	H43/31F	1980	Ex Red & White, 1993
944	HUD475S	Bristol VRT/SL3/6LXB	Eastern Coach Works	H43/31F	1977	Ex Oxford Bus Company, 1993
945	HUD480S	Bristol VRT/SL3/6LXB	Eastern Coach Works	H43/31F	1977	Ex Oxford Bus Company, 1993
946	HUD479S	Bristol VRT/SL3/6LXB	Eastern Coach Works	H43/31F	1977	Ex Oxford Bus Company, 1993
947	AET181T	Bristol VRT/SL3/6LXB	Eastern Coach Works	H43/31F	1979	Ex East Midland, 1994
948	VTV170S	Bristol VRT/SL3/6LXB	Eastern Coach Works	H43/31F	1978	Ex East Midland, 1994
949	DWF189V	Bristol VRT/SL3/6LXB	Eastern Coach Works	H43/31F	1980	Ex East Midland, 1994

Stagecoach have supported Showbus with their presence for some years. For the 1995 show Stagecoach Midland Red exhibited 912, B912ODU. The Leyland Olympian fleet consists of nine buses, of which 912 is numerically the last, and five dual-purpose vehicles allocated to express service duties. *Ralph Stevens*

958	WDA994T	Leyland Fleetline FE30AGR	MCW	H43/33F	1979	Ex West Midlands Travel, 1990
960	B960ODU	Leyland Olympian ONLXB/1R	Eastern Coach Works	DPH42/30F	1984	
961	B961ODU	Leyland Olympian ONLXB/1R	Eastern Coach Works	DPH42/30F	1984	
962	C962XVC	Leyland Olympian ONLXB/1RH	Eastern Coach Works	DPH42/29F	1985	
963	C963XVC	Leyland Olympian ONLXB/1RH	Eastern Coach Works	DPH42/29F	1985	
964	C964XVC	Leyland Olympian ONLXB/1RH	Eastern Coach Works	DPH42/29F	1985	

970-988 Leyland Atlantean AN68A/2R — Alexander AL — H49/37F — 1978-80 — Ex Busways, 1995-96

970	SCN252S	**974**	VCU301T	**978**	SCN276S	**982**	AVK174V	**986**	AVK169V
971	SCN253S	**975**	VCU310T	**979**	SCN281S	**983**	AVK168V	**987**	AVK140V
972	SCN265S	**976**	AVK172V	**980**	AVK181V	**984**	AVK182V	**988**	AVK145V
973	UVK298T	**977**	EJR106W	**981**	VCU304T	**985**	AVK167V		

1051	KIB8140	Leyland National 10351A/2R		B22DL	1978	Ex London Buses, 1991
1052	AIB4053	Leyland National 10351A/2R		B22DL	1978	Ex London Buses, 1991
1053	PIB8109	Leyland National 10351A/2R		B22DL	1978	Ex London Buses, 1991
4345	C102HKG	Ford Transit 190D	Robin Hood	B16F	1986	Ex Red & White, 1993
4364	C714FKE	Ford Transit 190D	Dormobile	B16F	1986	Ex East Kent, 1991

Previous Registrations:

3063VC	G543LWU	552OHU	A201RHT	A848VML	FRA64V
3273AC	BVP788V	6253VC	YDK917, JPU817, C472CAP	A8GGT	A202RHT
331HWD	BVP787V	9258VC	G554LWU	AIB4053	THX186S
3669DG	YKV811X	9737VC	C212PPE	CSV219	F846TLU
4012VC	KUB546V	9984PG	FYX815W	E315NWK	E422GAC, 6267AC
420GAC	CDG213Y	A6GGT	E630KCX	KIB8140	THX249S
4828VC	C211PPE	A75NAC	A190GVC, 420GAC	PIB8019	THX119S
498FYB	CDG207Y	A7GGT	B72OKV	WSU293	From New

SWANBROOK

Swanbrook Coaches Ltd; D J, K J & J A Thomas, Thomas House, St Margarets Road, Cheltenham, GL50 4DS

Depot :Pheasant Lane, Golden Valley, Staverton

KDF100P	Bedford YRT	Duple Dominant Express	DP57F	1975	
OJD151R	Leyland Fleetline FE30AGR	Park Royal	H44/29F	1976	Ex Stevensons, 1992
MVK546R	Leyland Atlantean AN68A/2R	Alexander AL	H48/34F	1977	Ex Colchester, 1990
MVK548R	Leyland Atlantean AN68A/2R	Alexander AL	H48/34F	1977	Ex Colchester, 1990
TDF103R	Bedford YMT	Plaxton Supreme III Express	C53F	1977	
ADD491S	Bedford YMT	Duple Dominant	C53F	1977	Ex Glos Transport Training, 1996
THX500S	Leyland Fleetline FE30ALR	Park Royal	H44/24D	1977	Ex Kinch, Barrow-on-Soar, 1994
THX340S	Leyland Fleetline FE30ALR	MCW	H44/24D	1978	Ex Kinch, Barrow-on-Soar, 1994
SDA566S	Leyland Fleetline FE30AGR	MCW	H44/33F	1978	Ex West Midlands Travel, 1992
SDA776S	Leyland Fleetline FE30AGR	MCW	H44/33F	1978	Ex West Midlands Travel, 1992
ADF106T	Bedford YMT	Plaxton Supreme III Express	C53F	1978	
XAK457T	Leyland National 11351A/1R		B52F	1978	Ex Lloyd, Nuneaton, 1994
WYV47T	Leyland Titan TNLXB2RR	Park Royal	H44/22D	1979	Ex Kinch, Barrow-on-Soar, 1995
FDD109T	Bedford YMT	Plaxton Supreme IV	C53F	1979	
JDG112V	Bedford YMT	Plaxton Supreme IV Express	C53F	1980	
MNW133V	Leyland National 2 NL116L11/1R		B52F	1980	Ex Kinch, Barrow-on-Soar, 1995
TIB6410	Leyland Tiger TRCTL11/3R	Plaxton Paramount 3200E	C57F	1983	Ex Cheltenham & Gloucester, 1993
TIB6411	Leyland Tiger TRCTL11/3R	Plaxton Paramount 3200E	C53F	1983	Ex Cheltenham & Gloucester, 1993
B120UUD	Leyland Tiger TRCTL11/3RH	Plaxton Paramount 3500 IIE	C51F	1985	Ex Oxford Bus Company, 1996
B121UUD	Leyland Tiger TRCTL11/3RH	Plaxton Paramount 3500 IIE	C51F	1985	Ex Oxford Bus Company, 1996
SIB4458	Volvo B10M-61	Plaxton Paramount 3500 II	C53F	1986	Ex Clarkes of London, 1992
D122EFH	Bedford YMT	Plaxton Derwent	B55F	1987	
D123EFH	Bedford YMT	Plaxton Derwent	B55F	1987	
A10SBK	Leyland Tiger TRCL10/3ARZM	Plaxton Paramount 3500 III	C49FT	1988	Ex Grimsby-Cleethorpes, 1994
A11SBK	Leyland Tiger TRCL10/3ARZM	Plaxton Paramount 3500 III	C49FT	1988	Ex Grimsby-Cleethorpes, 1994
G802FJX	Scania K93CRB	Van Hool Alizée	C55F	1990	Ex Abbeyways, Halifax, 1995
G804FJX	Scania K93CRB	Van Hool Alizée	C55F	1990	Ex Abbeyways, Halifax, 1995

Previous Registrations:

SIB4458	C179LWB	A10SBK	E510RFU, PS2743, E882HFW
TIB6410	A200RHT	A11SBK	E28RFU, PS3696, E881HFW
TIB6411	A214SAE	ADD491S	VDF518S, 6017WF

Livery: White, red and blue (older coaches); grey/silver, red, orange and yellow (newer coaches); yellow, orange and red (buses - but most carry overall adverts)

It is not normal policy to show overall adverts or rally pictures in the Bus Handbooks, though readers views on such matters would be appreciated. However, here are both with this picture of Swanbrook's MNW133V, a Leyland National 2 that was new to West Yorkshire before arriving with Swanbrook from Kinch in 1995.
Phillip Stephenson

Many Swanbrook vehicles have been bought in pairs and D122EFH is no exception. Photographed in Gloucester it is a Bedford YMT with Plaxton Derwent bodywork and one of the last from this General Motors subsidiary before production ceased. *Malc McDonald*

The latest coaches, again a pair, arrived in 1995 from Abbeyways in the form of two Van Hool Alizées. These have Scania K93 chassis with the smaller 280hp engine. Lettered for Swanbrook Holidays, this particular example was pictured while on an outing to the Royal Welsh Show. *David Donati*

TAPPINS

Tappins Coaches, Holiday House, Station Road, Didcot, Oxfordshire, OX11 7LZ

Depots : Station Road, Didcot and Collett Road Southmead Ind Est, Didcot.

770EWL	Leyland National 1151/1R/0401		B52F	1973	Ex South Wales, 1986
BFS14L	Leyland Atlantean AN68/1R	Alexander AL	O45/33F	1973	Ex Lothian, 1990
BFS34L	Leyland Atlantean AN68/1R	Alexander AL	O45/33F	1973	Ex Lothian, 1990
BFS48L	Leyland Atlantean AN68/1R	Alexander AL	O45/33F	1973	Ex Lothian, 1990
BFS49L	Leyland Atlantean AN68/1R	Alexander AL	O45/33F	1973	Ex Lothian, 1990
BFS50L	Leyland Atlantean AN68/1R	Alexander AL	O45/33F	1973	Ex Lothian, 1990
OSF939M	Leyland Atlantean AN68/1R	Alexander AL	O45/33F	1974	Ex Lothian, 1991
653GBU	Leyland National 2 NL116AL11/1R		B52F	1982	Ex AERE, Harwell, 1991
461XPB	Volvo B10M-61	Plaxton Viewmaster IV	C53F	1982	
500EFC	Volvo B10M-61	Plaxton Viewmaster IV	C53F	1982	
966MKE	Volvo B10M-61	Plaxton Supreme V	C53F	1982	
B161FWJ	Volvo B10M-61	Plaxton Paramount 3500	C53F	1985	
B163FWJ	Volvo B10M-61	Plaxton Paramount 3500	C53F	1985	
C323UFP	Volvo B10M-61	Plaxton Paramount 3500 II	C53F	1986	
C324UFP	Volvo B10M-61	Plaxton Paramount 3500 II	C53F	1986	
C325UFP	Volvo B10M-61	Plaxton Paramount 3500 II	C53F	1986	
YUE338	Volvo B10M-61	Plaxton Paramount 3500 II	C49FT	1986	
KBZ7145	Ford Transit 190	Carlyle	B16F	1986	Ex ?, 1994
D73HRU	Volvo B10M-61	Plaxton Paramount 3500 III	C53F	1987	
D74HRU	Volvo B10M-61	Plaxton Paramount 3500 III	C53F	1987	
D75HRU	Volvo B10M-61	Plaxton Paramount 3500 III	C53F	1987	
E471SON	MCW Metrobus DR102/63	MCW	H45/30F	1988	Ex London Buses, 1992
E257PEL	Toyota Coaster HB31R	Caetano Optimo	C19F	1988	
E258PEL	Toyota Coaster HB31R	Caetano Optimo	C19F	1988	

In addition to quality contracts and private hire work Tappins provide local services in the Didcot area and the opentop vehicles for The Oxford Classic Tour. *Opposite:* **Shown on duty is BFS49L, an Alexander-bodied Leyland Atlantean which was new to Edinburgh Corporation. Representing the coach fleet** *(below)* **is N179LHU, one of many Volvo B10Ms in the fleet that carry the new stylish TC lettering and livery.** *Phillip Stephenson*

E260PEL	Volvo B10M-61	Plaxton Paramount 3500 III	C53F	1988	
IIL1832	Aüwaerter Neoplan N122/3	Aüwaerter Skyliner	CH57/18DT	1988	Ex Voyager, Selby, 1992
F400DUG	Volvo B10M-60	Plaxton Paramount 3500 III	C48F	1989	Ex Wallace Arnold, 1993
F165XLJ	Volvo B10M-60	Plaxton Paramount 3500 III	C53F	1989	
F166XLJ	Volvo B10M-60	Plaxton Paramount 3500 III	C53F	1989	
G417VAY	Volvo B10M-60	Plaxton Paramount 3500 III	C53F	1990	
G418VAY	Volvo B10M-60	Plaxton Paramount 3500 III	C53F	1990	
G419VAY	Volvo B10M-60	Plaxton Paramount 3500 III	C53F	1990	
G504LWU	Volvo B10M-60	Plaxton Paramount 3500 III	C50F	1990	Ex Wallace Arnold, 1993
G505LWU	Volvo B10M-60	Plaxton Paramount 3500 III	C50F	1990	Ex Wallace Arnold, 1993
G506LWU	Volvo B10M-60	Plaxton Paramount 3500 III	C50F	1990	Ex Wallace Arnold, 1993
G507LWU	Volvo B10M-60	Plaxton Paramount 3500 III	C50F	1990	Ex Wallace Arnold, 1993
G508LWU	Volvo B10M-60	Plaxton Paramount 3500 III	C50F	1990	Ex Wallace Arnold, 1993
G509LWU	Volvo B10M-60	Plaxton Paramount 3500 III	C50F	1990	Ex Wallace Arnold, 1993
G510LWU	Volvo B10M-60	Plaxton Paramount 3500 III	C50F	1990	Ex Wallace Arnold, 1993
G511LWU	Volvo B10M-60	Plaxton Paramount 3500 III	C50F	1990	Ex Wallace Arnold, 1993
G512LWU	Volvo B10M-60	Plaxton Paramount 3500 III	C50F	1990	Ex Wallace Arnold, 1993
G513LWU	Volvo B10M-60	Plaxton Paramount 3500 III	C50F	1990	Ex Wallace Arnold, 1993
H261GRY	Volvo B10M-60	Plaxton Paramount 3500 III	C53F	1991	
H262GRY	Volvo B10M-60	Plaxton Paramount 3500 III	C53F	1991	
J854PUD	Dennis Dart 9.8SDL3012	Reeve Burgess Pointer	B43F	1992	
K301GDT	Volvo B10M-60	Van Hool Alizée	C53F	1993	
K302GDT	Volvo B10M-60	Van Hool Alizée	C53F	1993	
L540XJU	Mazda E2200	Howletts	M8	1993	
N171LHU	Volvo B10M-62	Van Hool Alizée	C53F	1996	
N172LHU	Volvo B10M-62	Van Hool Alizée	C53F	1996	
N173LHU	Volvo B10M-62	Van Hool Alizée	C53F	1996	
N174LHU	Volvo B10M-62	Van Hool Alizée	C53F	1996	
N175LHU	Volvo B10M-62	Van Hool Alizée	C53F	1996	
N176LHU	Volvo B10M-62	Van Hool Alizée	C53F	1996	
N177LHU	Volvo B10M-62	Van Hool Alizée	C53F	1996	
N178LHU	Volvo B10M-62	Van Hool Alizée	C53F	1996	
N179LHU	Volvo B10M-62	Van Hool Alizée	C53F	1996	
N180LHU	Volvo B10M-62	Van Hool Alizée	C53F	1996	

Previous Registrations:

461XPB	NBL904X	770EWL	NWN720M	KBZ7145	D826UTF
500EFC	NBL904X	966MKE	NBL906X	YUE338	C326UFP
653GBU	WBW735X	IIL1832	E480YWJ		

Livery: Orange, black and white

TEME VALLEY MOTORS

CA, FO & DB Griffiths, The Garage, Leintwardine, Herefordshire SY7 0JZ

JJX622N	Bedford YRQ	Plaxton Elite III Express	C45F	1975	Ex Evans, Tregaron, 1994
CFO902V	Bedford YLQ	Duple Dominant II Express	C45F	1980	
FUJ917W	Bedford YMT	Duple Dominant II	C53F	1980	Ex Whittle, Highley, 1985
FCJ400W	Bedford YMT	Duple Dominant II Express	C45F	1981	Ex Yeomans, Hereford, 1985
NJI5235	Bedford YNT	Plaxton Supreme VI Express	C53F	1982	Ex Go-Whittle, Kidderminster, 1994
KAB100X	Bedford YNT	Caetano Alpha	C53F	1982	
A644UNV	Mercedes-Benz L307D	Reeve Burgess	M12	1983	Ex Evans, Tregaron, 1991
A660ANT	Leyland Tiger TRCTL11/3R	Plaxton Paramount 3200E	C57F	1983	
TVM263	Leyland Tiger TRCTL11/3R	Jonckheere Jubilee P50	C53F	1983	Ex Barfoot, West End, 1994
D517OPP	Freight Rover Sherpa	Chassis Developments	C16F	1986	Ex Brownrigg, Egremont, 1993
D523GBX	Freight Rover Sherpa	Deansgate	C16F	1987	Ex Davies Bros, Pencader, 1990
D782AFO	Bedford CF	Dormobile	M12	1987	Ex Sargeants, Kington, 1994
E146TBO	MCW MetroRider MF150/78	MCW	B23F	1988	Ex Cardiff Bus, 1996
F720DAV	Freight Rover Sherpa	Freight Rover	M16	1989	Ex Stigwood, High Wycombe, 1994

Previous Registrations:

NJI5235 PNT849X, GBB254, MAB112X, XKH455, MUY121X
TVM263 BRN701Y, 8177VT, 9975VT, RJU94Y, 991FOT, NTP95Y

Livery: White and green

Leintwardine, close to the Shropshire border in the Teme Valley, is home to the Plaxton Supreme VI Express NJI5235. A college service to Hereford is operated on appropriate days. Bedfords still form the mainstay of the coach fleet, though two Leyland Tigers are also operated.

THAMES TRANSIT

Thames Transit Ltd, Horspath Road, Cowley, Oxfordshire, OX4 2RY

Depots : Horspath Road, Cowley; London Road, Chipping Norton and Corn Street, Witney.

1	L723JUD	Volvo B10M-60	Jonckheere Deauville 45	C49FT	1994	
2	L724JUD	Volvo B10M-60	Jonckheere Deauville 45	C49FT	1994	

3-7		Volvo B10M-60		Jonckheere Deauville 45		C49FT	1993		
3	L210GJO	**4**	L211GJO	**5**	L212GJO	**6**	L213GJO	**7**	L214GJO

8	N201CUD	Mercedes-Benz 711D	Marshall C19	DP28F	1995	
9	M103XBW	Volvo B10M-62	Berkhof Excellence	C51FT	1995	
10	H639UWR	Volvo B10M-60	Plaxton Paramount 3500 III	C48FT	1991	Ex Wallace Arnold, 1994
11	H640UWR	Volvo B10M-60	Plaxton Paramount 3500 III	C48FT	1991	Ex Wallace Arnold, 1994
12	N202CUD	Mercedes-Benz 711D	Marshall C19	DP28F	1995	
14	N203CUD	Mercedes-Benz 711D	Marshall C19	DP28F	1995	
15	H641UWR	Volvo B10M-60	Plaxton Paramount 3500 III	C48FT	1991	Ex Wallace Arnold, 1994
16	M104XBW	Volvo B10M-62	Berkhof Excellence	C51FT	1995	
17	H650UWR	Volvo B10M-60	Plaxton Paramount 3500 III	C48FT	1991	Ex Wallace Arnold, 1994
18	L159LBW	Volvo B10M-62	Jonckheere Deauville 45	C49FT	1994	
20	H914FTT	Volvo B10M-60	Ikarus Blue Danube	C49FT	1991	
21	L155LBW	Volvo B10M-62	Jonckheere Deauville 45	C49FT	1994	
22	H916FTT	Volvo B10M-60	Ikarus Blue Danube	C49FT	1991	
24	J499MOD	Volvo B10M-60	Ikarus Blue Danube	C49FT	1992	
25	M105XBW	Volvo B10M-62	Berkhof Excellence	C51FT	1995	
26	M106XBW	Volvo B10M-62	Berkhof Excellence	C51FT	1995	
27	H916PTG	Volvo B10M-60	Ikarus Blue Danube	C49FT	1991	Ex Hills of Tredegar, 1992
28	M107XBW	Volvo B10M-62	Berkhof Excellence	C42FT	1995	
29	L156LBW	Volvo B10M-62	Jonckheere Deauville 45	C49FT	1994	
30	L157LBW	Volvo B10M-62	Jonckheere Deauville 45	C49FT	1994	
31	L158LBW	Volvo B10M-62	Jonckheere Deauville 45	C49FT	1994	
32	N204CUD	Mercedes-Benz 711D	Marshall C19	DP28F	1995	
33	N205CUD	Mercedes-Benz 711D	Marshall C19	DP28F	1995	
34	N206CUD	Mercedes-Benz 711D	Marshall C19	DP28F	1995	

One of Thames Transit's principal routes is the 24-hour express service between Oxford and London operated as the Oxford Tube. All of the coaches used on this service are Volvo B10Ms, though three suppliers of bodywork feature. Shown here is the Jonckheere Deauville 45 which is produced in Roeselare, Belgium. Also used are Berkhof Excellence from Heerenveen in The Netherlands and Ikarus Blue Danube from Budapest in Hungary.
Richard Godfrey

Oxford city services developed by Thames Transit feature vehicles with dedicated liveries that emphasise the route branding. Shown here are three dual-doored Dennis Darts with Plaxton Pointer bodywork. ***Above*** **is 3049, N47EJO in purple and grey Cavalier livery while** ***(opposite)*** **are 3003, L712JUD in Blackbird Flyer scheme and 3026, M75VJO for the Rose Hill Runner service.** *Plaxton*

41-48	Volvo B10M-62	Berkhof Excellence 1000	C51FT	1996

41	N41MJO	**43**	N43MJO	**46**	N46MJO	**47**	N47MJO	**48**	N48MJO
42	N42MJO	**45**	N45MJO						

50	E829ATT	Mercedes-Benz 709D	Reeve Burgess Beaver	DP25F	1988	
64	F724FDV	Mercedes-Benz 709D	Reeve Burgess Beaver	B25F	1989	
74	F734FDV	Mercedes-Benz 709D	Reeve Burgess Beaver	B25F	1989	

104-139	Ford Transit VE6	Mellor	B16F	1986-87

104	D104PTT	**121**	D121PTT	**126**	D126PTT	**132**	D132PTT	**137**	D137PTT
107	D107PTT	**122**	D122PTT	**127**	D127PTT	**133**	D133PTT	**138**	D138PTT
108	D108PTT	**123**	D123PTT	**129**	D129PTT	**135**	D135PTT	**139**	D139PTT
109	D109PTT	**124**	D124PTT						

143	E826ATT	Ford Transit VE6	Mellor	B16F	1988	
200	LRV992	Leyland Titan PD2/12	Metro Cammell	O33/26R	1956	Ex Southdown, 1991
287	XTP287L	Leyland Atlantean AN68/1R	Alexander AL	H45/30D	1973	Ex Portsmouth, 1991

300-324	Mercedes-Benz 709D	Reeve Burgess Beaver	DP25F	1988

300	E300BWL	**305**	E305BWL	**310**	F310EJO	**315**	F315EJO	**320**	F320EJO
301	E301BWL	**306**	E306BWL	**311**	F311EJO	**316**	F316EJO	**321**	F321EJO
302	E302BWL	**307**	E307BWL	**312**	F312EJO	**317**	F317EJO	**322**	F322EJO
303	E303BWL	**308**	E308BWL	**313**	F313EJO	**318**	F318EJO	**323**	F323EJO
304	E304BWL	**309**	E309BWL	**314**	F314EJO	**319**	F319EJO	**324**	F324EJO

Cowley Rd., High St.
CITY CENTRE
1
Blackbird Leys Cowley City Centre
Blackbird Flyer
Thames Transit
L712 JUO
PLAXTON

ffley Rd., High St.
CITY CENTRE
3
Rose Hill · Iffley Road · City Centre
The Rose Hill Runner
Thames Transit
3026
M75 VJO
PLAXTON

Contravision is applied to the windows and used as part of Thames Transit's Park and Ride service livery. Seen here on the service is 2088, L948EOD, a Iveco TurboDaily 59-12 with Mellor dual-door bodywork. *Phillip Stephenson*

326-346		Mercedes-Benz 709D		Reeve Burgess Beaver		B25F	1989		
326	F776FDV	**329**	F766FDV	**332**	F769FDV	**339**	F409KOD	**345**	F403KOD
327	F764FDV	**330**	F767FDV	**333**	F770FDV	**344**	F402KOD	**346**	F746FDV
328	F765FDV	**331**	F768FDV						
347-354		Mercedes-Benz 709D		Carlyle		B29F	1990		
347	G947TDV	**349**	G949TDV	**351**	G951TDV	**353**	G843UDV	**354**	G954TDV
348	G948TDV	**350**	G950TDV	**352**	G952TDV				
355-366		Mercedes-Benz 811D		Carlyle		B33F	1990		
355	G831UDV	**358**	G834UDV	**361**	G837UDV	**363**	G839UDV	**365**	G841UDV
356	G832UDV	**359**	G835UDV	**362**	G838UDV	**364**	G840UDV	**366**	G842UDV
357	G833UDV	**360**	G836UDV						
638	D638NOD	Ford Transit 190D		Mellor		B16F	1987		
655	D655NOD	Ford Transit 190D		Mellor		B16F	1987		
901	N901PFC	Dennis Lance 11SDA3113		Plaxton Verde		B49F	1996		
902	N902PFC	Dennis Lance 11SDA3113		Plaxton Verde		B49F	1996		
903	N903PFC	Dennis Lance 11SDA3113		Plaxton Verde		B49F	1996		
997w	F280HOD	Leyland Tiger TRBTL11/2RP		Plaxton Derwent II		B54F	1988		
998w	F281HOD	Leyland Tiger TRBTL11/2RP		Plaxton Derwent II		B54F	1988		
999w	F282HOD	Leyland Tiger TRBTL11/2RP		Plaxton Derwent II		B54F	1988		
2017	K718UTT	Iveco Turbo Daily 59-12		Mellor Duet		B26D	1992/93		
2066	L318BOD	Iveco Turbo Daily 59-12		Mellor Duet		B26D	1993		
2069	L321BOD	Iveco Turbo Daily 59-12		Mellor Duet		B26D	1993		
2085-2090		Iveco Turbo Daily 59-12		Mellor Duet		B26D	1994		
2085	L945EOD	**2087**	L947EOD	**2088**	L948EOD	**2089**	L949EOD	**2090**	L950EOD
2086	L946EOD								

Over the years, the original service between Oxford and London has been widened to include towns on either side of the main corridor. The 390 service works to the south, calling at places such as Wallingford, Henley and Maidenhead. Dedicated to that service is 34, N206CUD, one of five Mercedes-Benz 711s placed in service in 1995. *Richard Godfrey*

3000-3013		Dennis Dart 9.8SDL3035		Plaxton Pointer		B37D		1994 (3006 is B39D)	
3000	L709JUD	**3003**	L712JUD	**3006**	L715JUD	**3009**	L718JUD	**3012**	L721JUD
3001	L710JUD	**3004**	L713JUD	**3007**	L716JUD	**3010**	L719JUD	**3013**	L722JUD
3002	L711JUD	**3005**	L714JUD	**3008**	L717JUD	**3011**	L720JUD		

3014-3050		Dennis Dart 9.8SDL3054		Plaxton Pointer		B37D		1995	
3014	M59VJO	**3022**	M69VJO	**3030**	M81WBW	**3037**	M89WBW	**3044**	M97WBW
3015	M61VJO	**3023**	M71VJO	**3031**	M82WBW	**3038**	M91WBW	**3045**	M98WBW
3016	M62VJO	**3024**	M73VJO	**3032**	M83WBW	**3039**	M92WBW	**3046**	M101WBW
3017	M63VJO	**3025**	M74VJO	**3033**	M84WBW	**3040**	M93WBW	**3047**	M102WBW
3018	M64VJO	**3026**	M75VJO	**3034**	M85WBW	**3041**	M94WBW	**3048**	M103WBW
3019	M65VJO	**3027**	M76VJO	**3035**	M86WBW	**3042**	M95WBW	**3049**	N47EJO
3020	M67VJO	**3028**	M78VJO	**3036**	M87WBW	**3043**	M96WBW	**3050**	N48EJO
3021	M68VJO	**3029**	M79VJO						

3051	N51KBW	Dennis Dart 9.8SDL3054	Plaxton Pointer	B37D	1996
3052	N52KBW	Dennis Dart 9.8SDL3054	Plaxton Pointer	B37D	1996
3053	N53KBW	Dennis Dart 9.8SDL3054	Plaxton Pointer	B37D	1996
3054	N54KBW	Dennis Dart 9.8SDL3054	Plaxton Pointer	B37D	1996

3055-3062		Dennis Dart 9.8SDL3054		Plaxton Pointer		B40F		1996	
3055	N56KBW	**3057**	N58KBW	**3059**	N61KBW	**3061**	N63KBW	**3062**	N64KBW
3056	N57KBW	**3058**	N59KBW	**3060**	N62KBW				

Previous Registrations:

Livery: Grey and blue (with several variations for local marketing schemes); grey and red (Oxford Tube coaches)

Named coaches: *1 New College; 2 Nuffield; 3 Merton; 4 Queens College; 5 Magdalen; 6 Pembroke; 7 St Catherine's College; 9 Manchester College; 16 Somerville; 18 Mansfield College; 19 Brasenose; 20 Balliol; 21 Christ Church Oxford; 25 Corpus Christi College; 26 Worcester College; 27 Wadham; 28 The Commuter; 29 Trinity; 30 Keble and 31 St Hilda's.*

Note: All vehicles were new to Transit Holdings companies except where shown.

WAINFLEET

Wainfleet Motor Services Ltd, Oaston Road, Nuneaton, Warwickshire, CV11 6JX

EVC211T	Bedford YMT	Duple Dominant II	C53F	1979	Ex Adams, Walsall, 1994
MIW5795	Volvo B10M-61	Plaxton Viewmaster IV	C53F	1981	Ex De Luxe, Mancetter, 1996
MIW5798	Bedford YNT	Plaxton Supreme IV	C53F	1982	Ex Bull, Tideswell, 1988
MIW5794	Volvo B10M-61	Jonckheere Jubilee	C53F	1983	Ex Longmynd, Pontesbury, 1996
MIW5793	Volvo B10M-61	Plaxton Paramount 3500 III	C50FT	1988	Ex Wallace Arnold, 1996
MIW5788	Leyland Tiger TRCTL11/3ARZ	Plaxton Paramount 3200 III	C53F	1988	Ex Park's, 1991
MIW5790	Volvo B10M-60	Plaxton Paramount 3500 III	C53F	1989	Ex Express Travel, 1996
MIW5787	Volvo B10M-60	Plaxton Paramount 3500 III	C53F	1989	Ex Stagecoach South (East Kent), 1996
MIW5786	Volvo B10M-60	Plaxton Paramount 3500 III	C53F	1989	Ex Stagecoach South (East Kent), 1996
MIW5785	Volvo B10M-60	Plaxton Paramount 3500 III	C53F	1990	Ex Wessex, 1995
MIW5796	Volvo B10M-60	Plaxton Paramount 3500 III	C49FT	1990	Ex Park's, 1996
MIW5797	Volvo B10M-60	Plaxton Paramount 3500 III	C49FT	1990	Ex Westerham Coaches, 1993
MIW5791	Volvo B10M-60	Plaxton Paramount 3500 III	C53F	1990	Ex Harry Shaw, 1994
MIW5789	Volvo B10M-60	Plaxton Paramount 3500 III	C53F	1990	Ex Silver Coach Lines, Edinburgh, 1994

Previous Registrations:

MIW5785	G562VHY	MIW5790	F227BHF	MIW5795	LBO11X
MIW5786	G907PKK	MIW5791	G84RGG	MIW5796	G45RGG
MIW5787	G908PKK	MIW5792		MIW5797	G82RGG
MIW5788	F203HSO	MIW5793	E305UUB	MIW5798	PWJ152X
MIW5789	G90RGG	MIW5794	UAB943Y		

A feature of the Wainfleet business is the use of MIW index marks. All except one vehicle now carry the marks which run consecutively. Pictured at Gloucester bus station is MIW5791 one of the 1990 intake of Plaxton-bodied Volvo B10Ms for Park's of Hamilton. *Robert Edworthy*

WESTWARD TRAVEL

MJ & LC Simmons, The Chipping, Kingswood, Wotton-under-Edge,
Gloucestershire GL12 8RT

Depot : near The Fox, Hawkesbury Upton.

	NOC595R	Leyland Fleetline FE30AGR	Park Royal	H43/33F	1976	Ex Beeston, Hadleigh, 1994
23	OHR190R	Leyland Fleetline FE30AGR	Eastern Coach Works	H43/31F	1977	Ex Stubbington, Benfleet, 1993
	URB161S	Bristol VRT/SL3/6LXB	Eastern Coach Works	H43/31F	1977	Ex Bugler, Bristol, 1995
	VPF287S	Bristol VRT/SL3/6LXB	Eastern Coach Works	H43/31F	1978	Ex Red Bus Services, Aylesbeare, 1995
	PTD673S	Leyland National 11351A/1R		B49F	1978	Ex GM Buses, 1987
	DAD2T	Ford R1114	Plaxton Supreme IV	C53F	1979	Ex Rover, Horsley, 1984
	WCU823T	Leyland Leopard PSU3E/4R	Plaxton Supreme IV Express	C53F	1979	Ex Wrays, Harrogate, 1987
	ERY790T	Ford R1114	Plaxton Supreme IV	C53F	1979	Ex Viking, Woodville, 1985
	ECS59V	Dennis Dominator DD120B	East Lancashire	H45/32F	1979	Ex A1 Service (McKinnon), 1988
	DPV881	Bova EL26/581	Bova Europa	C53F	1981	Ex Prestwood Travel, 1990
	VWX350X	Bova EL26/581	Bova Europa	C51F	1982	Ex Wallace Arnold, 1987
	HBH416Y	Leyland Tiger TRCTL11/3R	Plaxton Paramount 3200	C50F	1983	Ex Stevens, Bristol, 1991
	A930JOD	Leyland Tiger TRCTL11/3R	Plaxton Paramount 3200	C57F	1983	Ex Snell, Newton Abbot, 1992
	312XYB	DAF SB2300DHS585	Plaxton Paramount 3200	C55F	1984	Ex Rover, Horsley, 1989
	A661UHY	Bova FLD12.250	Bova Futura	C53F	1984	Ex Clayton, Leicester, 1995

Previous Registrations:

312XYB	A21KDF	DPV881	URW702X

Livery: Two-tone grey

Named Vehicles: 312XYB *Westward Consort;* DAD2T *Westward Invader;* DPV881 *Westward Duchess;* ERY790T *Westward Princess;* HBH416Y *Westward Ranger;* VWX350X *Westward Courier;* WCU823T *Westward Crusader;*

Five double-deck buses are operated by Westward Travel of Wootton-under-Edge. Below is Eastern Coach Works-bodied Daimler Fleetline OHR190R. *Robert Edworthy*

WOODSTONES COACHES

Woodstones Coaches Ltd, Arthur Drive, Hoo Farm Ind Est, Kidderminster
Worcestershire, DY11 7RA

F466WFX	Volvo B10M-60	Plaxton Paramount 3200 III	C57F	1989	Ex Smith, Tring, 1990
H633UWR	Volvo B10M-60	Plaxton Paramount 3500 III	C53F	1991	Ex Wallace Arnold, 1994
H634UWR	Volvo B10M-60	Plaxton Paramount 3500 III	C53FT	1991	Ex Wallace Arnold, 1995
H638UWR	Volvo B10M-60	Plaxton Paramount 3500 III	C53F	1991	Ex Ralph's, Langley, 1994
L48CNY	Volvo B10M-60	Plaxton Premiere 320	C57F	1993	Ex Bebb, Llantwit Fardre, 1996
M376MRU	Volvo B10M-61	Plaxton Premiere 320	C53F	1995	Ex Excelsior, Bournemouth, 1996

Previous Registrations:
M376MRU XEL24

Livery: White, red and yellow

WORTH'S

Worths Motor Services Ltd, The Garage, Enstone, Oxfordshire, OX7 4LQ

Depots : The Garage, Enstone and Burford Road, Chipping Norton.

OJD401R	Leyland Fleetline FE30AGR	Park Royal	H44/24D	1977	Ex Tappins, Didcot, 1993
UVO125S	Leyland Leopard PSU3E/4R	Duple Dominant II Express	C49F	1977	Ex Williams, Oadby, 1996
XKV488S	Ford R1114	Plaxton Supreme III Express	C53F	1978	Ex Pathfinder, Freckleton, 1985
GNF7V	Leyland Titan TNLXB/1RF	Park Royal	H47/26F	1979	Ex Thames Transit, 1989
TDM770V	Volvo B58-61	Duple Dominant II	C57F	1980	Ex Lofty's, Bridge Trafford, 1986
FFC322V	Volvo B58-61	Plaxton Supreme IV	C57F	1980	
FUD322W	Volvo B58-61	Plaxton Supreme IV	C57F	1980	
JWL322W	Volvo B58-61	Plaxton Supreme IV	C57F	1980	
SKG406Y	Volvo B10M-61	Plaxton Paramount 3200	C57F	1983	Ex K&P John, Llanharry, 1990
A112MUD	Leyland Tiger TRCTL11/3RH	Plaxton Paramount 3200 E	C51F	1984	Ex The Oxford Bus Co, 1996
A114MUD	Leyland Tiger TRCTL11/3RH	Plaxton Paramount 3200 E	C51F	1984	Ex The Oxford Bus Co, 1996
A116PBW	Leyland Tiger TRCTL11/3RH	Plaxton Paramount 3200 E	C51F	1984	Ex The Oxford Bus Co, 1996
A118PBW	Leyland Tiger TRCTL11/3RH	Plaxton Paramount 3200 E	C50F	1984	Ex The Oxford Bus Co, 1996
B123UUD	Leyland Tiger TRCTL11/3RH	Plaxton Paramount 3200 E	C51F	1984	Ex The Oxford, Bus Co 1996
B124UUD	Leyland Tiger TRCTL11/3RH	Plaxton Paramount 3200 E	C51F	1984	Ex The Oxford Bus Co, 1996
551DJB	Volvo B10M-61	Plaxton Paramount 3500 II	C53F	1986	Ex Wallace Arnold, 1991
774YPG	Volvo B10M-61	Plaxton Paramount 3200 III	C49F	1988	
XSK144	Volvo B10M-61	Plaxton Paramount 3200 III	C57F	1988	
KAZ2755	Volvo B10M-61	Plaxton Paramount 3200 III	C57F	1989	
F464WFX	Volvo B10M-60	Plaxton Paramount 3200 III	C55F	1989	Ex Independent, Horsforth, 1994
G144MNH	Volvo B10M-60	Jonckheere Deauville P599	C51FT	1990	Ex Harris, Catshill, 1995
G844VAY	Volvo B10M-60	Duple 320	C57F	1989	Ex Crawford, Neilston, 1992
H443JLJ	Volvo B10M-60	Plaxton Paramount 3200 III	C57F	1990	Ex Bere Regis & District, 1993

Previous Registrations:

551DJB	C121CWR	KAZ2755	F322MFC
774YPG	F318GWL	XSK144	F396HFC

Livery: Silver and blue

Worths Motor Services have recently disposed of its two London Buses Leyland Titans. The remaining vehicle is from the same batch as the Cheltenham & Gloucester units which were new to SELNEC pte. This vehicle is seen at the Enstone depot during spring 1996. Most of the fleet is now coach based, with Plaxton-bodied Volvo B58's regularly used on contract services. Pictured on such duties in Oxford is FUD322V, a 12 metre example which has been with the company since new.
Tony Wilson/Richard Godfrey

YARRANTON

Yarranton Brothers Ltd, Eardiston, Tenbury Wells, Worcestershire, WR15 8JL

w	KYA99N	Bedford YRT	Duple Dominant	C53F	1975	Ex Beeline, Warminster, 1993
	MSF738P	Bedford YRT	Alexander AYS	B53F	1976	Ex Dean Forest, Joys Green, 1989
w	OHY791R	Bedford YMT	Duple Dominant	C53F	1977	Ex Applegate, Newport, 1994
	VJA660S	Bedford YMT	Plaxton Supreme III	C53F	1978	Ex Evans, Tregaron, 1995
	WVJ300T	Bedford YMT	Plaxton Supreme IV Express	C53F	1978	Ex Yeomans, Hereford, 1984
	EUH573V	Bedford YLQ	Plaxton Supreme IV	C35F	1980	Ex Evans, Tregaron, 1989
	PIB5773	Bedford YNT	Duple Dominant	C53F	1982	
	ALJ915A	Mercedes-Benz 0303/15R	Mercedes-Benz	C53F	1983	Ex Young, Romsley, 1991
	C21KBM	Bedford YNV Venturer	Plaxton Paramount 3200 II	C53F	1986	Ex Bedford demonstrator, 1987
	D684TPP	Freight Rover Sherpa	Chassis Developments	M16	1987	
	E242FJU	Mercedes-Benz 609D	Reeve Burgess	C19F	1987	
	RDZ4287	Volvo B10M-61	Jonckheere Jubilee P599	C53F	1988	Ex Tellings-Golden Miller, Cardiff, 1992
	G474BJU	Mercedes-Benz 0303	Plaxton Paramount 3500 III	C48FT	1990	Ex Abbeyways, Halifax, 1995
	J461LLK	Sanos S311-21	Sanos Charisma	C35FT	1992	Ex Hamilton of Uxbridge, 1995

Previous Registrations:

ALJ915A	PUL81Y	PIB5773	HFO547X
G474BJU	G561MWX, SIB1294	RDZ4287	E506KNV

This shortened Bedford YLQ is seen operating with Yarrantons at Kidderminster. The Plaxton Supreme IV body seats only 35 passengers although a standard 10-metre version carried typically forty-five and may be considered the forerunner of the current midibus fashion. The Duple body for the YLQ is shown opposite. *Martin Grosberg*

YEOMANS

Yeomans Canyon Travel Ltd, 21/3 Three Elms Trading Estate, Hereford HR4 9PO

1	PGR619N	Bedford YRT	Willowbrook 001	B53F	1974	Ex Jolly, South Hylton, 1980
2	VWK8S	Bedford YMT	Plaxton Supreme III	C53F	1978	Ex Freeman, Uffington, 1988
3	C795WVJ	Bedford YNT	Plaxton Paramount 3200 II	C53F	1986	
4	A700OCJ	Bedford YNT	Plaxton Paramount 3200 E	C53F	1983	
5	JVJ439P	Bedford YRQ	Duple Dominant	B50F	1975	
6	KVJ700Y	Bedford YNT	Plaxton Paramount 3200 E	C53F	1982	
7	C858XCJ	Bedford YMP	Plaxton Paramount 3200 II	C45F	1986	
8	YOI2747	Volvo B10M-61	Plaxton Paramount 3500 III	C49FT	1988	Ex Snow, Great Wakering, 1994
9	J74CVJ	Dennis Dart 8.5SDL3003	Alexander Dash	B30F	1991	
10	A569TNV	Bedford YMT	Wadham Stringer Vanguard	B53F	1983	Ex Evans, Tregaron, 1994
11	FAD442Y	Bedford YNT	Plaxton Paramount 3200	C46F	1983	
12	H554YCJ	Dennis Dart 9SDL3002	Carlyle Dartline	B36F	1991	
14	F702MCJ	Dennis Javelin 12SDA1907	Duple 320	C57F	1989	
15	PTV596X	Bedford YNT	Plaxton Supreme IV Express	C53F	1982	Ex Evans, Tregaron, 1989
16	YOI298	Volvo B10M-61	Plaxton Paramount 3500 III	C49FT	1988	Ex Empress, Bethnal Green, 1993
17	B904TCJ	Bedford YNT	Plaxton Paramount 3200 II	C53F	1984	
18	BGR631W	Bedford YMT	Duple Dominant	B55F	1981	Ex Davies Bros, Carmarthen, 1989
19	GHL192L	Bristol VRT/SL2/6LX	Eastern Coach Works	H43/31F	1972	Ex Countryman, Ibstock, 1989
20	MKP181W	Bedford YMT	Wadham Stringer Vanguard	B61F	1981	Ex Cave, Shirley, 1994
21	FAD257Y	Bedford YNT	Plaxton Paramount 3200	C42FT	1983	
22	YCT502	DAF MB200DKFL600	Caetano Algarve	C53F	1984	Ex Buchanan, Stretton Sugwas, 1991
23	WUG143S	Bedford YMT	Duple Dominant II	C53F	1978	Ex Buchanan, Stretton Sugwas, 1991
24	PAB911T	Bedford YLQ	Duple Dominant II	C31FT	1979	Ex Buchanan, Stretton Sugwas, 1991
25	H465MCC	Dennis Dart 9.8SDL3004	Carlyle Dartline	B43F	1990	Ex Padarn, Caernarfon, 1995
27	PCJ900R	Bedford YLQ	Plaxton Supreme III	C45F	1977	

Duple also bodied Bedford YLQs, including the specially shortened chassis which filled the gap between the 29-seat front engined PJK and the normal 10-metre 45-seater. Shown operating with Yeomans, PAB911T with Duple Dominant II bodywork which was the only type used for the YLQ. It was preparing for a journey to Fownhope when photographed. *Robert Edworthy*

Yeomans provide several services from Hereford with town services departing from the Tesco (city) bus station while rural connections are made from the country bus station near to the rail station. Interesting vehicles used on local services are Bedford YMQs, and 37, YLW894X carries a Lex Maxeta body. It is seen in the bus station adjacent to Tesco, the terminus of service 117 to Newton Farm. The only double-deck bus in Yeomans fleet is a Bristol VRT used exclusively for school duties. GHL192L carries one of the early Eastern Coach Works bodies with BET-style windscreens and was new to West Riding in 1972. *Martin Grosberg*

This early Dennis Dart, H457MEY was previously with north Wales operator, Padarn. It now sports Yeomans full livery and is one of five Darts in the fleet to three different lengths. *Richard Eversden*

28	BNO699T	Bedford YRT	Duple Dominant II	C53F	1979	Ex Buchanan, Stretton Sugwas, 1991
29	YXI2755	Bedford YLQ	Duple Dominant	B47F	1978	Ex AEE, Winfrith, 1993
30	CCG550V	Bedford YMT	Duple Dominant	B61F	1980	Ex Yorkshire Rider, 1993
31	YCV834	Volvo B10M-61	Plaxton Paramount 3500 III	C53F	1988	Ex Flights, 1992
32	PFO300R	Bedford YLQ	Plaxton Supreme III	C45F	1977	
33	NNT522P	Bedford YRT	Willowbrook 001	B60F	1976	Ex Brown, Telford, 1978
34	J158CCJ	Dennis Javelin 12SDA1919	Berkhof Excellence 2000	C53F	1991	
35	WVJ530T	Bedford YLQ	Duple Dominant II Express	C45F	1979	
36	K436GVJ	Renault Master T35D	Cymric	C16F	1992	
37	YLW894X	Bedford YMQ	Lex Maxeta	B35F	1982	Ex R & I, Milton Keynes, 1993
38	A300RCJ	Bedford YNT	Plaxton Paramount 3200	C53F	1984	
39	H457MEY	Dennis Dart 9.8SDL3004	Carlyle Dartline	B43F	1991	Ex Padarn, Caernarfon, 1995
40	L482ADG	Ford Transit VE6	Ford	M8	1994	Ex private owner, 1996
41	HCJ909X	Bedford YNT	Duple Dominant	B63F	1982	
42	FCY294W	Bedford YMQ	Duple Dominant	DP45F	1980	Ex Western Scottish, 1996
43	EVJ300W	Bedford YMT	Plaxton Supreme IV Express	C53F	1981	
44	EFO300W	Bedford YMT	Duple Dominant	B60F	1980	
45	FCY283W	Bedford YMQ	Duple Dominant	DP45F	1980	Ex Cave, Shirley, 1995
46	KCJ200Y	Bedford YMQ	Plaxton Supreme V Express	C45F	1982	
47	G212TDV	Volkswagen Microbus	Devon Conversions	M8	1989	
48	G249RTT	Volkswagen Microbus	Devon Conversions	M7	1989	
49	FCY295W	Bedford YMQ	Duple Dominant	B47F	1981	Ex County, 1995
53	FCY280W	Bedford YMQ	Duple Dominant	DP45F	1980	Ex Cave, Shirley, 1995
54	LXI9357	Volvo B10M-61	Plaxton Paramount 3500 III	C49FT	1988	Ex Daisy, Broughton, 1994
55	M341SCJ	Volvo B10M-62	Plaxton Expressliner 2	C46FT	1995	
56	M342SCJ	Volvo B10M-62	Plaxton Expressliner 2	C44FT	1995	
57	M343SCJ	Volvo B10M-62	Plaxton Expressliner 2	C46FT	1995	
58	J249SOC	Dennis Dart 9.8SDL3004	Carlyle Dartline	B40F	1991	Ex Cave, Shirley, 1995
	JKG32W	Bedford YMQ	Duple Dominant	B47F	1981	Ex Evans, Tregaron, 1996

Previous Registrations:

FAD257Y	LFO400Y, YCV834	YLW894X	LCY300X, ULL897, GGK239X,
FAD442Y	KVJ789Y, YCT502	YLW896X	LCY298X, 33LUG, GGK238X, RIB4316
LXI9357	E589UHS	YOI298	E277VJW, PJI1830, Exxxxxx
PAB911T	AUJ742T, WLT642	YOI2747	E577VDA, FTG9, A2GUK
YCT502	A644WCY, 278TNY, A654XWN	YXI2755	YRY988T
YCV834	E477VDA		

Livery: Cream and green with orange on most coaches; white (National Express) 55-7.

Index to vehicles

Registration	Operator
BVP818V	S'coach Midland Red
BWE196T	N N Cresswell
C21KBM	Yarrington
C48HKK	N N Cresswell
C102HKG	S'coach Midland Red
C108HKG	Kestrel
C114PUJ	Chiltern Queens
C125DWR	Springs Tours
C175HYD	N N Cresswell
C301PNP	Midland Red West
C304PNP	Midland Red West
C305PNP	Midland Red West
C306PNP	Midland Red West
C307PNP	Midland Red West
C308PNP	Midland Red West
C309PNP	Midland Red West
C310PNP	Midland Red West
C311PNP	Midland Red West
C312PNP	Midland Red West
C313PNP	Midland Red West
C314PNP	Midland Red West
C315OFL	H & H Motors
C315PNP	Midland Red West
C316PNP	Midland Red West
C317PNP	Midland Red West
C318PNP	Midland Red West
C319PNP	Midland Red West
C319UFP	Dudley's Coaches
C320PNP	Midland Red West
C321PNP	Midland Red West
C322PNP	Midland Red West
C323PNP	Midland Red West
C323UFP	Tappins
C324PNP	Midland Red West
C324UFP	Tappins
C325PNP	Midland Red West
C325UFP	Tappins
C326PNP	Midland Red West
C327PNP	Midland Red West
C328PNP	Midland Red West
C329PNP	Midland Red West
C330PNP	Midland Red West
C331PNP	Midland Red West
C332PNP	Midland Red West
C333PNP	Midland Red West
C334PNP	Midland Red West
C335PNP	Midland Red West
C336PNP	Midland Red West
C337PNP	Midland Red West
C338PNP	Midland Red West
C339PNP	Midland Red West
C340PNP	Midland Red West
C341PNP	Midland Red West
C342PNP	Midland Red West
C343PNP	Midland Red West
C344PNP	Midland Red West
C345PNP	Midland Red West
C346PNP	Midland Red West
C347PNP	Midland Red West
C348PNP	Midland Red West
C349PNP	Midland Red West
C350PNP	Midland Red West
C351PNP	Midland Red West
C352PNP	Midland Red West
C353PNP	Midland Red West
C354FBO	Cheney Coaches
C354PNP	Midland Red West
C355ALJ	Newbury Coaches
C355PNP	Midland Red West
C356PNP	Midland Red West
C357PNP	Midland Red West
C358PNP	Midland Red West
C359PNP	Midland Red West
C360PNP	Midland Red West
C361RUY	Midland Red West
C362RUY	Midland Red West
C363RUY	Midland Red West
C364RUY	Midland Red West
C365RUY	Midland Red West
C366RUY	Midland Red West
C367RUY	Midland Red West
C368RUY	Midland Red West
C369RUY	Midland Red West
C370RUY	Midland Red West
C371RUY	Midland Red West
C372RUY	Midland Red West
C373RUY	Midland Red West
C374RUY	Midland Red West
C375RUY	Midland Red West
C376RUY	Midland Red West
C378PCD	Charlton Services
C379PCD	Charlton Services
C380KUX	Hollands
C381RUY	Midland Red West
C382RUY	Midland Red West
C385RUY	Midland Red West
C386RUY	Midland Red West
C387RUY	Midland Red West
C388RUY	Midland Red West
C389RUY	Midland Red West
C390RUY	Midland Red West
C391RUY	Midland Red West
C392RUY	Midland Red West
C393RUY	Midland Red West
C394RUY	Midland Red West
C395RUY	Midland Red West
C396RUY	Midland Red West
C397RUY	Midland Red West
C398RUY	Midland Red West
C399RUY	Midland Red West
C400RUY	Midland Red West
C401RUY	Midland Red West
C402RUY	Midland Red West
C402XFO	Newbury Coaches
C403RUY	Midland Red West
C404RUY	Midland Red West
C438SJU	Andy James
C452CWR	Marchants
C455CWR	Marchants
C474CAP	Cottrell's
C475BHY	Midland Red West
C476BHY	Midland Red West
C477BHY	Midland Red West
C480BFB	Cheltenham & G
C483BHY	Midland Red West
C487BHY	Midland Red West
C488BHY	Midland Red West
C490BHY	Midland Red West
C491BHY	Midland Red West
C492BHY	Midland Red West
C497BHY	Midland Red West
C498BHY	Midland Red West
C499BFB	Cheltenham & G
C499BHY	Midland Red West
C516DND	Springs Tours
C531TJF	Bromyard Omnibus
C581SHC	Midland Red West
C582SHC	Midland Red West
C583SHC	Midland Red West
C584SHC	Midland Red West
C585SHC	Midland Red West
C586SHC	Midland Red West
C587SHC	Midland Red West
C588SHC	Midland Red West
C589SHC	Midland Red West
C617SFH	Cheltenham & G
C621SFH	Cheltenham & G
C626SFH	Cheltenham & G
C631SFH	Cheltenham & G
C633SFH	Cheltenham & G
C636SFH	Cheltenham & G
C637SFH	Cheltenham & G
C639SFH	Cheltenham & G
C640SFH	Cheltenham & G
C641SFH	Cheltenham & G
C642SFH	Cheltenham & G
C643SFH	Cheltenham & G
C644SJM	Chiltern Queens
C645SFH	Cheltenham & G
C651XDF	Cheltenham & G
C652XDF	McLeans
C659XDF	Cheltenham & G
C685MWJ	Dudley's Coaches
C693VAD	Cheltenham & G
C694VAD	Cheltenham & G
C696VAD	Cheltenham & G
C697VAD	Cheltenham & G
C705FKE	S'coach Midland Red
C714FKE	S'coach Midland Red
C724DHW	Sargeants
C738CUC	Cheltenham & G
C788FRL	Midland Red West
C790FRL	Midland Red West
C795WVJ	Yeomans
C858XCJ	Yeomans
C899REG	Alexcars
C900JGA	Sargeants
C918YBF	Hollands
C962XVC	S'coach Midland Red
C963XVC	S'coach Midland Red
C964XVC	S'coach Midland Red
C975HOX	Rogers of Martley
C976HOX	Rogers of Martley
C977HOX	Rogers of Martley
C985HOX	Midland Red West
C986HOX	Midland Red West
C987HOX	Midland Red West
CBV16S	S'coach Midland Red
CCG550V	Yeomans
CDG213Y	S'coach Midland Red
CFO902V	Teme Valley
CFX320T	Barry's Coaches
CHL772	Guide Friday
CIB7615	Sargeants
CIB9321	Sargeants
CIL3526	Barry's Coaches
CIL9223	Dudley's Coaches
CIW9129	Hardings
CRO671K	Marchants
CRU184C	Guide Friday
CSU243	Charlton Services
CSU432	Charlton Services
CTX397V	Boomerang Bus Co
CUD219Y	The Oxford Bus Co
CUD220Y	The Oxford Bus Co
CUD221Y	The Oxford Bus Co
CUD222Y	The Oxford Bus Co
CUD223Y	The Oxford Bus Co
CUD224Y	The Oxford Bus Co
CWF733T	Guide Friday
CWF736T	Guide Friday
CWG743V	Guide Friday
CWG744V	Guide Friday
CWG763V	Guide Friday
D27UCW	Alexcars
D34ENH	Chiltern Queens
D34KAX	S'coach Midland Red
D43KAX	S'coach Midland Red
D45KAX	S'coach Midland Red
D47KAX	S'coach Midland Red
D67OVP	Andy James
D73HRU	Tappins
D74HRU	Tappins
D75HRU	Tappins
D83WWV	Smith's
D104PTT	Thames Transit
D105TFT	Redline
D107PTT	Thames Transit
D108PTT	Thames Transit
D109PTT	Thames Transit
D118TFT	Hollands
D121PTT	Thames Transit
D122EFH	Swanbrook
D122PTT	Thames Transit
D123EFH	Swanbrook
D123PTT	Thames Transit
D124HMT	Smith's
D124PTT	Thames Transit
D126PTT	Thames Transit
D127PTT	Thames Transit
D129PTT	Thames Transit
D132PTT	Thames Transit
D133NON	Hollands
D133PTT	Thames Transit
D135PTT	Thames Transit
D137PTT	Thames Transit
D138PTT	Thames Transit
D139PTT	Thames Transit
D158BPH	Hardings
D160UGA	Cottrell's
D166RAK	Bromyard Omnibus
D225LWY	Dudley's Coaches
D247OOJ	Hollands
D262HFX	Chiltern Queens
D271OOJ	S'coach Midland Red
D290XCX	Bennetts
D291ALR	Chauffeurs
D392KND	Hollands
D396LRL	Kestrel
D470DWP	Hardings
D504NWG	Chiltern Queens
D506NWG	Chiltern Queens
D517OPP	Teme Valley
D523GBX	Teme Valley
D581VBV	Cheltenham & G
D604HTC	Cheltenham & G
D605HTC	Cheltenham & G
D606HTC	Cheltenham & G
D638NOD	Thames Transit
D655NOD	Thames Transit
D684TPP	Yarrington
D700BJF	Smith's
D708WEY	Kestrel
D723JUB	Hollands
D735OOG	S'coach Midland Red
D736OOG	S'coach Midland Red
D750SJO	The Oxford Bus Co
D763KWT	Midland Red West
D782AFO	Teme Valley
D803NBO	Cottrell's

Registration	Operator
D810PUK	Soudley Valley
D822UTF	The Oxford Bus Co
D823UTF	The Oxford Bus Co
D824UTF	The Oxford Bus Co
D827PUK	N N Cresswell
D829KWT	Sargeants
D851CKV	S'coach Midland Red
D852CKV	S'coach Midland Red
D856CKV	S'coach Midland Red
D857CKV	S'coach Midland Red
D858CKV	S'coach Midland Red
D862CKV	S'coach Midland Red
D869BDG	Ken Rose
D881BDF	Bennetts
D885CKV	S'coach Midland Red
D888MJA	McLeans
D917GRU	Bromyard Omnibus
D957WJH	Ken Rose
D969PJA	Bromyard Omnibus
D974PJW	Johnsons
D978PJA	Springs Tours
DAD2T	Westward Travel
DDD200T	Pulham's
DFB222W	JBC-Malvernian
DFC884R	Heyfordian
DHA986Y	Primrose Motors
DHW350W	Cheltenham & G
DHW352W	Cheltenham & G
DJS203	Lugg Valley
DNE545Y	Cheltenham & G
DPV881	Westward Travel
DRB60T	Springs Tours
DRB61T	De Luxe
DSD977V	Cheltenham & G
DSK514	Hardings
DSK515	Hardings
DSK516	Hardings
DSK593	Hardings
DSK594	Hardings
DSU114	Astons
DUP143S	James Bevan
DWF189V	S'coach Midland Red
DWF194V	S'coach Midland Red
DWF195V	S'coach Midland Red
DWJ564V	Guide Friday
DWK407T	Newbury Coaches
DWU839H	Guide Friday
E26JBD	Astons
E61SUH	Grayline
E62EVJ	Rogers of Martley
E62SUH	Boomerang Bus Co
E65EVJ	Lugg Valley
E68SUH	Grayline
E95RWR	Sargeants
E102OUH	Cheltenham & G
E130YUD	The Oxford Bus Co
E131YUD	The Oxford Bus Co
E132KYW	Dukes Coaches
E132YUD	The Oxford Bus Co
E133YUD	The Oxford Bus Co
E134YUD	The Oxford Bus Co
E137KYW	Dukes Coaches
E146TBO	Teme Valley
E164TWO	Boomerang Bus Co
E168TWO	Sargeants
E170TWO	Boomerang Bus Co
E175TWO	Boomerang Bus Co
E175TWW	Cheney Coaches
E200YTM	Kestrel
E215PWY	Grayline
E225CFC	The Oxford Bus Co
E226CFC	The Oxford Bus Co
E227CFC	The Oxford Bus Co
E228CFC	The Oxford Bus Co
E229CFC	The Oxford Bus Co
E242FJU	Yarrington
E245CGA	Astons
E257PEL	Tappins
E258PEL	Tappins
E260PEL	Tappins
E300BWL	Thames Transit
E301BWL	Thames Transit
E302BWL	Thames Transit
E303BWL	Thames Transit
E304BWL	Thames Transit
E305BWL	Thames Transit
E306BWL	Thames Transit
E307BWL	Thames Transit
E308BWL	Thames Transit
E309BWL	Thames Transit
E315NWK	S'coach Midland Red
E318UUB	K W Beard Ltd
E322UUB	Geoff Willetts
E323UUB	James Bevan
E346AAM	Barry's Coaches
E388FLD	N N Cresswell
E403WAM	N N Cresswell
E405MPX	Smith's
E406HAB	Midland Red West
E407HAB	Midland Red West
E408HAB	Midland Red West
E409HAB	Midland Red West
E410HAB	Midland Red West
E411HAB	Midland Red West
E412KUY	Midland Red West
E413KUY	Midland Red West
E414KUY	Midland Red West
E415KUY	Midland Red West
E416KUY	Midland Red West
E417KUY	Midland Red West
E418KUY	Midland Red West
E419KUY	Midland Red West
E420KUY	Midland Red West
E421KUY	Midland Red West
E422KUY	Midland Red West
E423KUY	Midland Red West
E424KUY	Midland Red West
E425KUY	Midland Red West
E426KUY	Midland Red West
E427KUY	Midland Red West
E428KUY	Midland Red West
E429KUY	Midland Red West
E430KUY	Midland Red West
E431KUY	Midland Red West
E432KUY	Midland Red West
E433KUY	Midland Red West
E433YHL	S'coach Midland Red
E434KUY	Midland Red West
E435KUY	Midland Red West
E436KUY	Midland Red West
E437KUY	Midland Red West
E438KUY	Midland Red West
E439KUY	Midland Red West
E471SON	Tappins
E478AFJ	Cottrell's
E533PRU	Chiltern Queens
E559UHS	Lugg Valley
E663JAD	Cheltenham & G
E665JAD	Cheltenham & G
E667JAD	Cheltenham & G
E675UNE	Dudley's Coaches
E676KDG	Cheltenham & G
E688UNE	Cottrell's
E701GNH	Cheney Coaches
E735VWJ	James Bevan
E751VJO	The Oxford Bus Co
E752VJO	The Oxford Bus Co
E753VJO	The Oxford Bus Co
E754VJO	The Oxford Bus Co
E755HJF	Newbury Coaches
E755VJO	The Oxford Bus Co
E756VJO	The Oxford Bus Co
E757VJO	The Oxford Bus Co
E758XWL	The Oxford Bus Co
E759XWL	The Oxford Bus Co
E760XWL	The Oxford Bus Co
E761XWL	The Oxford Bus Co
E762XWL	The Oxford Bus Co
E826ATT	Thames Transit
E829ATT	Thames Transit
E830EUT	Newbury Coaches
E833EUT	Newbury Coaches
E833LNP	Springs Tours
E903VWG	Kestrel
E988NMK	Smith's
EAD122T	Cottrell's
EAH890Y	Midland Red West
EAP989V	The Oxford Bus Co
EAP999V	The Oxford Bus Co
EBM440T	Charlton Services
EBW106Y	Chiltern Queens
EBW107Y	Chiltern Queens
EBZ8205	James Bevan
ECS59V	Westward Travel
ECY987V	James Bevan
EDF269T	Charlton Services
EDG250L	Castleways
EFO300W	Yeomans
EJR106W	S'coach Midland Red
EOI4376	Grayline
EPM137V	Barry's Coaches
ERV250D	Guide Friday
ERV254D	Guide Friday
ERY790T	Westward Travel
ESU940	Heyfordian
ETO161L	Guide Friday
EUH573V	Yarrington
EVC211T	Wainfleet
EVJ300W	Yeomans
EWS740W	Cheltenham & G
EWS743W	Cheltenham & G
EWS746W	Cheltenham & G
EWS748W	Cheltenham & G
EWS751W	Cheltenham & G
EWW208T	De Luxe
F56SAD	Hollands
F57YBO	Andy James
F66SMC	Cottrell's
F71LAL	S'coach Midland Red
F112TEE	Astons
F115TWP	Go Whittle
F116TWP	Go Whittle
F135LJO	The Oxford Bus Co
F136LJO	The Oxford Bus Co
F137LJO	The Oxford Bus Co
F138LJO	The Oxford Bus Co
F139LJO	The Oxford Bus Co
F139TWP	Go Whittle
F146UFH	Ken Rose
F151TWP	Go Whittle
F152TWP	Go Whittle
F165XLJ	Tappins
F166XLJ	Tappins
F167UDG	K W Beard Ltd
F183UFH	Cottrell's
F244SAB	Go Whittle
F254MGB	Barry's Coaches
F274JWL	McLeans
F280HOD	Thames Transit
F281HOD	Thames Transit
F282HOD	Thames Transit
F308LBW	McLeans
F309RMH	Cottrell's
F310EJO	Thames Transit
F311DET	Cheltenham & G
F311EJO	Thames Transit
F312EJO	Thames Transit
F313EJO	Thames Transit
F314EJO	Thames Transit
F315EJO	Thames Transit
F316EJO	Thames Transit
F317EJO	Thames Transit
F318EJO	Thames Transit
F318EWF	Bennetts
F319EJO	Thames Transit
F320EJO	Thames Transit
F321EJO	Thames Transit
F322EJO	Thames Transit
F323EJO	Thames Transit
F324BRN	Sargeants
F324EJO	Thames Transit
F344TSC	Chiltern Queens
F358GBW	McLeans
F366MUT	Lewis's
F387KVJ	N N Cresswell
F400DUG	Tappins
F402KOD	Thames Transit
F403KOD	Thames Transit
F409KOD	Thames Transit
F426RRY	Chauffeurs
F434ENB	Chauffeurs
F464WFX	Worth's
F466WFX	Woodstones Coaches
F501ANY	The Oxford Bus Co
F502ANY	The Oxford Bus Co
F503ANY	The Oxford Bus Co
F504ANY	The Oxford Bus Co
F505CBO	The Oxford Bus Co
F556NJM	The Oxford Bus Co
F557NJM	The Oxford Bus Co
F558NJM	The Oxford Bus Co
F559NJM	The Oxford Bus Co
F560NJM	The Oxford Bus Co
F607JSS	Bennetts
F622SAY	Lewis's
F623SAY	Lewis's
F624SAY	Lewis's
F634FNA	Pulham's
F639HVM	Chauffeurs
F642KCX	Bennetts
F643OHD	Bennetts
F660EDH	Sargeants
F660PWK	S'coach Midland Red
F660RTL	Marchants
F661PWK	S'coach Midland Red
F677PDF	Cheltenham & G
F683UDP	Chauffeurs
F700PAY	Barry's Coaches
F702MCJ	Yeomans
F715RDG	K W Beard Ltd

Registration	Operator
F720DAV	Teme Valley
F724FDV	Thames Transit
F734FDV	Thames Transit
F746FDV	Thames Transit
F752JLG	Astons
F763LBW	The Oxford Bus Co
F764FDV	Thames Transit
F765FDV	Thames Transit
F766FDV	Thames Transit
F767FDV	Thames Transit
F768FDV	Thames Transit
F769FDV	Thames Transit
F770FDV	Thames Transit
F776FDV	Thames Transit
F791GNA	Newbury Coaches
F803KCJ	Newbury Coaches
F830GKO	Rogers of Martley
F833LCJ	Primrose Motors
F834LCJ	Primrose Motors
F846TLU	S'coach Midland Red
F850NJO	McLeans
F864PAC	S'coach Midland Red
F868TNH	Springs Tours
F871UAC	S'coach Midland Red
F872UAC	S'coach Midland Red
F886SRT	James Bevan
F892JHA	Johnsons
F903RWP	Go Whittle
F904RWP	Go Whittle
F905YWY	Boomerang Bus Co
F907RWP	Go Whittle
F915YWY	Boomerang Bus Co
F940WFA	Dudley's Coaches
F944KTA	Astons
F952ENH	Dudley's Coaches
F986TTF	Chiltern Queens
F992HGE	Lugg Valley
FAD257Y	Yeomans
FAD442Y	Yeomans
FAD708T	Castleways
FBZ7356	Heyfordian
FBZ7357	Heyfordian
FCJ400W	Teme Valley
FCY280W	Yeomans
FCY283W	Yeomans
FCY294W	Yeomans
FCY295W	Yeomans
FDD109T	Swanbrook
FDF965	Pulham's
FDV804V	Nichols
FEH1Y	Midland Red West
FFC322V	Worth's
FFJ473V	JBC-Malvernian
FIL7662	Heyfordian
FIL7664	Heyfordian
FIL8317	Heyfordian
FIL8441	Heyfordian
FIL8446	Heyfordian
FNP98W	Andy James
FSU803	Dukes Coaches
FUD322W	Worth's
FUJ917W	Teme Valley
FUJ918V	Go Whittle
FUJ924V	Go Whittle
FUJ938V	N N Cresswell
FUJ940V	Go Whittle
FUJ950V	N N Cresswell
FWL778Y	The Oxford Bus Co
FWL779Y	The Oxford Bus Co
FWL781Y	The Oxford Bus Co
FWT956J	Guide Friday
G22UWL	Cheney Coaches
G26XBK	S'coach Midland Red
G29HDW	Go Whittle
G42RGG	Chauffeurs
G57RGG	Lugg Valley
G82BHP	Castleways
G99XUD	McLeans
G101AAD	Cheltenham & G
G101HNP	Midland Red West
G102AAD	Cheltenham & G
G102HNP	Midland Red West
G102JNP	Go Whittle
G103AAD	Cheltenham & G
G103HNP	Midland Red West
G104AAD	Cheltenham & G
G104HNP	Midland Red West
G105AAD	Cheltenham & G
G105HNP	Midland Red West
G106HNP	Midland Red West
G107HNP	Midland Red West
G108HNP	Midland Red West
G108JNP	Go Whittle
G109HNP	Midland Red West
G110HNP	Midland Red West
G111HNP	Midland Red West
G111JNP	Go Whittle
G112HNP	Midland Red West
G113HNP	Midland Red West
G114HNP	Midland Red West
G115HNP	Midland Red West
G115OGA	S'coach Midland Red
G116HNP	Midland Red West
G117HNP	Midland Red West
G118HNP	Midland Red West
G119HNP	Midland Red West
G120HNP	Midland Red West
G121HNP	Midland Red West
G122HNP	Midland Red West
G123HNP	Midland Red West
G124HNP	Midland Red West
G125HNP	Midland Red West
G126HNP	Midland Red West
G127HNP	Midland Red West
G128HNP	Midland Red West
G129HNP	Midland Red West
G130HNP	Midland Red West
G130JNP	Go Whittle
G131HNP	Midland Red West
G131JNP	Go Whittle
G132HNP	Midland Red West
G133HNP	Midland Red West
G134GOL	Astons
G134HNP	Midland Red West
G135HNP	Midland Red West
G136HNP	Midland Red West
G137HNP	Midland Red West
G138HNP	Midland Red West
G139HNP	Midland Red West
G140GOJ	Alexcars
G140HNP	Midland Red West
G141HNP	Midland Red West
G142HNP	Midland Red West
G143HNP	Midland Red West
G144HNP	Midland Red West
G144MNH	Worth's
G145HNP	Midland Red West
G146HNP	Midland Red West
G147HNP	Midland Red West
G148HNP	Midland Red West
G149HNP	Midland Red West
G150HNP	Midland Red West
G171XDX	Johnsons
G185JWP	Hardings
G186JWP	Hardings
G212TDV	Yeomans
G222KWE	The Oxford Bus Co
G224EOA	Redline
G230VWL	The Oxford Bus Co
G231VWL	The Oxford Bus Co
G232VWL	The Oxford Bus Co
G233VWL	The Oxford Bus Co
G234VWL	The Oxford Bus Co
G235VWL	The Oxford Bus Co
G249RTT	Yeomans
G271UFB	Chauffeurs
G290XFH	Geoff Willetts
G301WHP	S'coach Midland Red
G302WHP	S'coach Midland Red
G303WHP	S'coach Midland Red
G327SVV	Newbury Coaches
G382RCW	Bennetts
G417VAY	Tappins
G418VAY	Tappins
G419VAY	Tappins
G420WFP	Johnsons
G422YAY	Lewis's
G448CDG	Marchants
G474BJU	Yarrington
G490PNF	Astons
G504LWU	Tappins
G505LWU	Tappins
G506LWU	Tappins
G507LWU	Tappins
G508LWU	Tappins
G509LWU	Tappins
G510LWU	Tappins
G511LWU	Tappins
G512LWU	Tappins
G513LWU	Tappins
G518LWU	Newbury Coaches
G528LWU	S'coach Midland Red
G529LWU	S'coach Midland Red
G530LWU	S'coach Midland Red
G531LWU	S'coach Midland Red
G532LWU	S'coach Midland Red
G533LWU	Cheltenham & G
G534LWU	Cheltenham & G
G535LWU	S'coach Midland Red
G546LWU	Cheltenham & G
G547LWU	Cheltenham & G
G548LWU	Cheltenham & G
G564LWX	McLeans
G577RNC	Astons
G621XLO	The Oxford Bus Co
G659TCJ	Primrose Motors
G679AAD	Cheltenham & G
G680AAD	Cheltenham & G
G680YLP	Pulham's
G681AAD	Cheltenham & G
G682AAD	Cheltenham & G
G683AAD	Cheltenham & G
G684AAD	Cheltenham & G
G711VRY	Cheney Coaches
G769WFC	The Oxford Bus Co
G770WFC	The Oxford Bus Co
G771WFC	The Oxford Bus Co
G772WFC	The Oxford Bus Co
G773WFC	The Oxford Bus Co
G774WFC	The Oxford Bus Co
G775WFC	The Oxford Bus Co
G776WFC	The Oxford Bus Co
G777WFC	The Oxford Bus Co
G778WFC	The Oxford Bus Co
G779WFC	The Oxford Bus Co
G780WFC	The Oxford Bus Co
G781WFC	The Oxford Bus Co
G782WFC	The Oxford Bus Co
G783WFC	The Oxford Bus Co
G802FJX	Swanbrook
G804FJX	Swanbrook
G831UDV	Thames Transit
G832UDV	Thames Transit
G833UDV	Thames Transit
G834UDV	Thames Transit
G835UDV	Thames Transit
G836UDV	Thames Transit
G837UDV	Thames Transit
G838UDV	Thames Transit
G839UDV	Thames Transit
G840UDV	Thames Transit
G841UDV	Thames Transit
G842UDV	Thames Transit
G843UDV	Thames Transit
G844VAY	Worth's
G865BPD	Redline
G905WAY	Bennetts
G947TDV	Thames Transit
G948TDV	Thames Transit
G949TDV	Thames Transit
G950TDV	Thames Transit
G951TDV	Thames Transit
G952TDV	Thames Transit
G954TDV	Thames Transit
G965PRC	Astons
G970WNR	Lewis's
G975KJX	Bennetts
G993DDF	Marchants
G997OVA	Johnsons
G999KJX	Kestrel
GAZ8573	Charlton Services
GBB254	Go Whittle
GBB997N	Soudley Valley
GBB999N	Soudley Valley
GBU2V	Cottrell's
GBU6V	Cottrell's
GDF332V	Cottrell's
GDF650L	Marchants
GGD664T	Astons
GGM110W	The Oxford Bus Co
GHL191L	Guide Friday
GHL192L	Yeomans
GIL1481	Smith's
GIL2782	Rogers of Martley
GIL4267	Barry's Coaches
GJI7173	Charlton Services
GNF6V	Cheltenham & G
GNF8V	Cheltenham & G
GNF9V	Cheltenham & G
GNF10V	Cheltenham & G
GNF11V	Cheltenham & G
GNF7V	Worth's
GOI7376	Catteralls
GOL413N	S'coach Midland Red
GOL426N	S'coach Midland Red
GPA624V	H & H Motors
GPX546X	Chauffeurs
GRC887N	Guide Friday
GRC890N	Guide Friday
GSC858T	Astons

Registration	Operator
GTO333N	Guide Friday
GTO334N	Guide Friday
GTO335N	Guide Friday
GTP95X	Andy James
GTP97X	Andy James
GTX746W	S'coach Midland Red
GTX754W	S'coach Midland Red
GUD708L	Charlton Services
GVO715N	Guide Friday
GVO721N	Johnsons
GWV926V	Guide Friday
GYE280W	The Oxford Bus Co
GYS896D	Guide Friday
H2HCR	Hardings
H3HCR	Hardings
H4HCR	Hardings
H5HCR	Hardings
H11PSV	Bennetts
H17GOW	Go Whittle
H19GOW	Go Whittle
H39UNH	Alexcars
H64XBD	James Bevan
H180EJF	Johnsons
H189RWF	The Oxford Bus Co
H191RWF	The Oxford Bus Co
H261GRY	Tappins
H262GRY	Tappins
H345KDF	Pulham's
H383HFH	Castleways
H401MRW	S'coach Midland Red
H402MRW	S'coach Midland Red
H403MRW	S'coach Midland Red
H404CJF	Chauffeurs
H404MRW	S'coach Midland Red
H406MRW	S'coach Midland Red
H443JLJ	Worth's
H457MEY	Yeomans
H465MCC	Yeomans
H495MRW	S'coach Midland Red
H554YCJ	Yeomans
H562FLE	Cheney Coaches
H569GMO	H & H Motors
H633UWR	Woodstones Coaches
H634HBW	Pearces
H634UWR	Woodstones Coaches
H638UWR	Woodstones Coaches
H639UWR	Thames Transit
H640UWR	Thames Transit
H641UWR	Thames Transit
H650MBF	Chauffeurs
H650UWR	Thames Transit
H726UKY	Astons
H753ELP	Sargeants
H787JFC	Cheney Coaches
H788RWJ	Chiltern Queens
H808RWJ	McLeans
H902AHS	Lugg Valley
H912XGA	S'coach Midland Red
H914FTT	Thames Transit
H916FTT	Thames Transit
H916PTG	Thames Transit
H918SCX	Sargeants
H937DRJ	Geoff Willetts
HAX331W	Bennetts
HBH416Y	Westward Travel
HCJ909X	Yeomans
HCR233	Hardings
HCR601	Hardings
HDF661	Pulham's
HEU122N	S'coach Midland Red
HFG451T	Carterton
HHU838X	Sargeants
HIL2295	Heyfordian
HIL4995	Kestrel
HIL6075	Cheltenham & G
HIL6649	McLeans
HIL7404	Chauffeurs
HIL7772	Alexcars
HIL8516	Catteralls
HJB451W	The Oxford Bus Co
HJB452W	The Oxford Bus Co
HJB453W	The Oxford Bus Co
HJB454W	The Oxford Bus Co
HJB455W	The Oxford Bus Co
HJB456W	The Oxford Bus Co
HJB457W	The Oxford Bus Co
HJB458W	The Oxford Bus Co
HJB459W	The Oxford Bus Co
HJB461W	The Oxford Bus Co
HJB462W	The Oxford Bus Co
HJI531	Astons
HKL826	Guide Friday
HKL836	Guide Friday
HNU670R	Cheltenham & G
HSV720	Heyfordian
HUD475S	S'coach Midland Red
HUD479S	S'coach Midland Red
HUD480S	S'coach Midland Red
HVN601N	N N Cresswell
HYY3	Astons
IIB145	Astons
IIL1237	Barry's Coaches
IIL1361	Primrose Motors
IIL1832	Tappins
J2HCR	Hardings
J2NNC	N N Cresswell
J3HCR	Hardings
J21GCX	Bennetts
J74CVJ	Yeomans
J100OFC	Pearces
J13OVA	Johnsons
J140NJO	The Oxford Bus Co
J141NJO	The Oxford Bus Co
J158CCJ	Yeomans
J201RAC	Lewis's
J249SOC	Yeomans
J304THP	S'coach Midland Red
J305THP	S'coach Midland Red
J306KFP	Castleways
J336UHP	Astons
J362BNW	Castleways
J407PRW	S'coach Midland Red
J408PRW	S'coach Midland Red
J409PRW	S'coach Midland Red
J410PRW	S'coach Midland Red
J411PRW	S'coach Midland Red
J412PRW	S'coach Midland Red
J413PRW	S'coach Midland Red
J414PRW	S'coach Midland Red
J415PRW	S'coach Midland Red
J416PRW	S'coach Midland Red
J417PRW	S'coach Midland Red
J418PRW	S'coach Midland Red
J461LLK	Yarrington
J499MOD	Thames Transit
J625KUT	Chauffeurs
J663CVJ	Newbury Coaches
J670LGA	Heyfordian
J687LGA	Heyfordian
J688MFE	Castleways
J689LGA	Heyfordian
J689MFE	Castleways
J718KBC	Lewis's
J719KBC	Lewis's
J733USF	Smith's
J854PUD	Tappins
J914MDG	Pulham's
JAL876N	Guide Friday
JDB950V	Barry's Coaches
JDE973X	Springs Tours
JDG112V	Swanbrook
JDK911P	Bromyard Omnibus
JDX574V	Cheney Coaches
JEY124Y	Marchants
JHU899X	Cheltenham & G
JHU912X	Cheltenham & G
JIL3755	Chauffeurs
JIL3756	Chauffeurs
JIL3757	Chauffeurs
JIL8323	Chauffeurs
JJX622N	Teme Valley
JKG32W	Yeomans
JKV430V	Johnsons
JND258V	Grayline
JNJ24V	Springs Tours
JOI4693	Grayline
JOU160P	Cheltenham & G
JOV766P	Barry's Coaches
JOX502P	S'coach Midland Red
JOX503P	S'coach Midland Red
JOX504P	S'coach Midland Red
JOX505P	Cheltenham & G
JPY505	Go Whittle
JRB663V	N N Cresswell
JUD597W	Pearces
JVJ439P	Yeomans
JWB847W	Dudley's Coaches
JWL322W	Worth's
JWV127W	The Oxford Bus Co
JWV128W	The Oxford Bus Co
K2BCC	Bennetts
K2HCR	Hardings
K2SUP	Alexcars
K3HCR	Hardings
K4HCR	Hardings
K6GOW	Go Whittle
K30ARJ	Andy James
K33GOW	Go Whittle
K100OMP	Pearces
K105UFP	Grayline
K196SFH	Rogers of Martley
K200OMP	Pearces
K201GRW	Alexcars
K208SFP	Chauffeurs
K301GDT	Tappins
K302GDT	Tappins
K306ARW	S'coach Midland Red
K308YKG	Cheltenham & G
K373HHK	Cheney Coaches
K420ARW	S'coach Midland Red
K421ARW	S'coach Midland Red
K422ARW	S'coach Midland Red
K423ARW	S'coach Midland Red
K424ARW	S'coach Midland Red
K425ARW	S'coach Midland Red
K436AAV	Hollands
K436GVJ	Yeomans
K441ATF	Ken Rose
K521EFL	S'coach Midland Red
K522EFL	S'coach Midland Red
K523EFL	S'coach Midland Red
K524EFL	S'coach Midland Red
K525EFL	S'coach Midland Red
K526EFL	S'coach Midland Red
K527EFL	S'coach Midland Red
K528EFL	S'coach Midland Red
K529EFL	S'coach Midland Red
K530EFL	S'coach Midland Red
K544OGA	Astons
K718UTT	Thames Transit
K750UJO	The Oxford Bus Co
K751UJO	The Oxford Bus Co
K752UJO	The Oxford Bus Co
K753UJO	The Oxford Bus Co
K754UJO	The Oxford Bus Co
K755UJO	The Oxford Bus Co
K801OMW	Cheltenham & G
K802OMW	Cheltenham & G
K830HVJ	Newbury Coaches
K929VDF	Pulham's
KAB100X	Teme Valley
KAD351V	Barry's Coaches
KAZ2755	Worth's
KBC6V	Grayline
KBZ7145	Tappins
KCJ200Y	Yeomans
KDF100P	Swanbrook
KEY212P	Sargeants
KFO570P	JBC-Malvernian
KGE299Y	Hollands
KHC813K	Guide Friday
KHC814K	Guide Friday
KHC815K	Guide Friday
KHC817K	Guide Friday
KHH376W	Cheltenham & G
KHL460W	Dudley's Coaches
KHT122P	S'coach Midland Red
KHT124P	S'coach Midland Red
KIB7026	Grayline
KIB7027	Grayline
KIB8140	S'coach Midland Red
KKW65P	Cheltenham & G
KPT583T	N N Cresswell
KSU839P	Guide Friday
KTL25V	Marchants
KTL26V	Marchants
KUB667V	Guide Friday
KUX211W	Sargeants
KUY98X	Dudley's Coaches
KVC385V	Lewis's
KVC386V	Lewis's
KVJ700Y	Yeomans
KWP111X	Hardings
KYA99N	Yarrington
KYN291X	The Oxford Bus Co
KYN308X	The Oxford Bus Co
KYV300W	The Oxford Bus Co
KYV317X	The Oxford Bus Co
KYV328X	The Oxford Bus Co
KYV370X	The Oxford Bus Co
KYV381X	The Oxford Bus Co
KYV392X	The Oxford Bus Co
KYV452X	The Oxford Bus Co
KYV457X	The Oxford Bus Co
KYV493X	The Oxford Bus Co
KYV510X	The Oxford Bus Co
KYV516X	The Oxford Bus Co
KYV519X	The Oxford Bus Co
KYV524X	The Oxford Bus Co
KYV530X	The Oxford Bus Co

Registration	Operator
L2HCR	Hardings
L2NNC	N N Cresswell
L11VWL	Pearces
L26CAY	Heyfordian
L48CNY	Woodstones Coaches
L91WBX	Sargeants
L111RBT	Barry's Coaches
L113YAB	Redline
L120CUY	Go Whittle
L150HUD	The Oxford Bus Co
L151HUD	The Oxford Bus Co
L152HUD	The Oxford Bus Co
L153HUD	The Oxford Bus Co
L154HUD	The Oxford Bus Co
L155HUD	The Oxford Bus Co
L155LBW	Thames Transit
L156LBW	Thames Transit
L157LBW	Thames Transit
L158LBW	Thames Transit
L159LBW	Thames Transit
L201AAB	Midland Red West
L202AAB	Midland Red West
L203AAB	Midland Red West
L204AAB	Midland Red West
L205AAB	Midland Red West
L206AAB	Midland Red West
L207AAB	Midland Red West
L208AAB	Midland Red West
L209AAB	Midland Red West
L210AAB	Midland Red West
L210GJO	Thames Transit
L211AAB	Midland Red West
L211GJO	Thames Transit
L212AAB	Midland Red West
L212GJO	Thames Transit
L213AAB	Midland Red West
L213GJO	Thames Transit
L214AAB	Midland Red West
L214GJO	Thames Transit
L215AAB	Midland Red West
L216AAB	Midland Red West
L217AAB	Midland Red West
L218AAB	Midland Red West
L219AAB	Midland Red West
L220AAB	Midland Red West
L221AAB	Midland Red West
L223AAB	Midland Red West
L224AAB	Midland Red West
L225AAB	Midland Red West
L226AAB	Midland Red West
L227AAB	Midland Red West
L228AAB	Midland Red West
L229AAB	Midland Red West
L230AAB	Midland Red West
L230BUT	Astons
L231AAB	Midland Red West
L232AAB	Midland Red West
L233AAB	Midland Red West
L234AAB	Midland Red West
L235AAB	Midland Red West
L236AAB	Midland Red West
L237AAB	Midland Red West
L248CCK	Cheltenham & G
L307SKV	S'coach Midland Red
L308YDU	S'coach Midland Red
L309YDU	S'coach Midland Red
L310YDU	S'coach Midland Red
L310YDU	S'coach Midland Red
L311YDU	S'coach Midland Red
L312YDU	S'coach Midland Red
L313YDU	S'coach Midland Red
L314YDU	S'coach Midland Red
L315YDU	S'coach Midland Red
L316YDU	S'coach Midland Red
L317YDU	S'coach Midland Red
L318BOD	Thames Transit
L318YDU	S'coach Midland Red
L319YDU	S'coach Midland Red
L321BOD	Thames Transit
L321YDU	S'coach Midland Red
L322AAB	Midland Red West
L322YDU	S'coach Midland Red
L323YDU	S'coach Midland Red
L324YDU	S'coach Midland Red
L325YDU	S'coach Midland Red
L326YKV	S'coach Midland Red
L327YKV	S'coach Midland Red
L328YKV	S'coach Midland Red
L329YKV	S'coach Midland Red
L330CHB	Cheltenham & G
L330YKV	S'coach Midland Red
L348MKU	Grayline
L349MKU	Catteralls
L353MKU	Geoff Willetts
L360YNR	Astons
L424ANP	Newbury Coaches
L451YAC	S'coach Midland Red
L452YAC	S'coach Midland Red
L453YAC	S'coach Midland Red
L454YAC	S'coach Midland Red
L455YAC	S'coach Midland Red
L456YAC	S'coach Midland Red
L463RDN	Bennetts
L482ADG	Yeomans
L535XUT	Heyfordian
L540XJU	Tappins
L549OWC	Johnsons
L584EPC	Johnsons
L686CDD	Cheltenham & G
L687CDD	Cheltenham & G
L688CDD	Cheltenham & G
L689CDD	Cheltenham & G
L690CDD	Cheltenham & G
L691CDD	Cheltenham & G
L692CDD	Cheltenham & G
L693CDD	Cheltenham & G
L694CDD	Cheltenham & G
L695CDD	Cheltenham & G
L696CDD	Cheltenham & G
L709FWO	Cheltenham & G
L709JUD	Thames Transit
L710FWO	Cheltenham & G
L710JUD	Thames Transit
L711FWO	Cheltenham & G
L711JUD	Thames Transit
L712FWO	Cheltenham & G
L712JUD	Thames Transit
L713JUD	Thames Transit
L714JUD	Thames Transit
L715JUD	Thames Transit
L716JUD	Thames Transit
L717JUD	Thames Transit
L718JUD	Thames Transit
L719JUD	Thames Transit
L720JUD	Thames Transit
L721JUD	Thames Transit
L722JUD	Thames Transit
L723JUD	Thames Transit
L724JUD	Thames Transit
L740YGE	Heyfordian
L743YGE	Heyfordian
L745YGE	Heyfordian
L801HJO	The Oxford Bus Co
L802HJO	The Oxford Bus Co
L803HJO	The Oxford Bus Co
L803XDG	Cheltenham & G
L804HJO	The Oxford Bus Co
L804XDG	Cheltenham & G
L805XDG	Cheltenham & G
L806XDG	Cheltenham & G
L831CDG	Cheltenham & G
L832CDG	Cheltenham & G
L833CDG	Cheltenham & G
L834CDG	Cheltenham & G
L835CDG	Cheltenham & G
L836CDG	Cheltenham & G
L837CDG	Cheltenham & G
L838CDG	Cheltenham & G
L839CDG	Cheltenham & G
L840CDG	Cheltenham & G
L841CDG	Cheltenham & G
L842CDG	Cheltenham & G
L945EOD	Thames Transit
L946EOD	Thames Transit
L947EOD	Thames Transit
L948EOD	Thames Transit
L949EOD	Thames Transit
L950EOD	Thames Transit
LBO10X	De Luxe
LBU781V	DRM
LCJ626Y	N N Cresswell
LDD488	Pulham's
LDS381V	De Luxe
LDZ2502	Heyfordian
LDZ2503	Heyfordian
LDZ3145	Carterton
LFB681P	JBC-Malvernian
LFH675V	Barry's Coaches
LFH719V	Castleways
LFH720V	Castleways
LFR860X	Cheltenham & G
LFR861X	Cheltenham & G
LFR873X	Cheltenham & G
LGB855V	Marchants
LHT724P	S'coach Midland Red
LHT725P	S'coach Midland Red
LHT726P	Cheltenham & G
LIB1797	N N Cresswell
LIL2050	Catteralls
LIL2665	Charlton Services
LIL3065	Boomerang Bus Co
LIL3066	Boomerang Bus Co
LIL9267	Boomerang Bus Co
LIL9268	Boomerang Bus Co
LIL9270	Boomerang Bus Co
LIL9968	K W Beard Ltd
LMS157W	Sargeants
LOA832X	Midland Red West
LOA838X	Cheltenham & G
LPY459W	Bennetts
LRV992	Thames Transit
LSV548	Go Whittle
LUL511X	Cheltenham & G
LVS433V	Primrose Motors
LWS33Y	Cheltenham & G
LWS34Y	Cheltenham & G
LWS35Y	Cheltenham & G
LWS36Y	Cheltenham & G
LWS37Y	Cheltenham & G
LWS38Y	Cheltenham & G
LWS39Y	Cheltenham & G
LWS40Y	Cheltenham & G
LWS41Y	Cheltenham & G
LXI9357	Yeomans
M2HCR	Hardings
M3HCR	Hardings
M10JMJ	Johnsons
M10RGJ	Johnsons
M30ARJ	Andy James
M59VJO	Thames Transit
M61VJO	Thames Transit
M62VJO	Thames Transit
M63VJO	Thames Transit
M64VJO	Thames Transit
M65VJO	Thames Transit
M67VJO	Thames Transit
M68VJO	Thames Transit
M69VJO	Thames Transit
M71VJO	Thames Transit
M73VJO	Thames Transit
M74VJO	Thames Transit
M75VJO	Thames Transit
M76VJO	Thames Transit
M78VJO	Thames Transit
M79VJO	Thames Transit
M81WBW	Thames Transit
M82WBW	Thames Transit
M83WBW	Thames Transit
M84WBW	Thames Transit
M85WBW	Thames Transit
M86WBW	Thames Transit
M87WBW	Thames Transit
M89WBW	Thames Transit
M91WBW	Thames Transit
M92WBW	Thames Transit
M93WBW	Thames Transit
M94WBW	Thames Transit
M95WBW	Thames Transit
M96WBW	Thames Transit
M97WBW	Thames Transit
M98WBW	Thames Transit
M101WBW	Thames Transit
M102WBW	Thames Transit
M103WBW	Thames Transit
M103XBW	Thames Transit
M104XBW	Thames Transit
M105XBW	Thames Transit
M106XBW	Thames Transit
M107XBW	Thames Transit
M139LNP	Ken Rose
M151KDD	Castleways
M173SBT	Johnsons
M201LHP	S'coach Midland Red
M202LHP	S'coach Midland Red
M203LHP	S'coach Midland Red
M204LHP	S'coach Midland Red
M205LHP	S'coach Midland Red
M209LHP	S'coach Midland Red
M210LHP	S'coach Midland Red
M219TCJ	Newbury Coaches
M238MRW	Midland Red West
M239MRW	Midland Red West
M240MRW	Midland Red West
M241MRW	Midland Red West
M242MRW	Midland Red West
M243MRW	Midland Red West
M244MRW	Midland Red West
M245MRW	Midland Red West
M246MRW	Midland Red West
M247MRW	Midland Red West

Registration	Operator
M248MRW	Midland Red West
M249MRW	Midland Red West
M250MRW	Midland Red West
M251MRW	Midland Red West
M252MRW	Midland Red West
M253MRW	Midland Red West
M254MRW	Midland Red West
M255MRW	Midland Red West
M256MRW	Midland Red West
M300ARJ	Andy James
M331LHP	S'coach Midland Red
M332LHP	S'coach Midland Red
M334LHP	S'coach Midland Red
M335LHP	S'coach Midland Red
M336LHP	S'coach Midland Red
M337LHP	S'coach Midland Red
M338LHP	S'coach Midland Red
M339LHP	S'coach Midland Red
M340LHP	S'coach Midland Red
M341LHP	S'coach Midland Red
M341SCJ	Yeomans
M342LHP	S'coach Midland Red
M342SCJ	Yeomans
M343LHP	S'coach Midland Red
M343SCJ	Yeomans
M344LHP	S'coach Midland Red
M345LHP	S'coach Midland Red
M346LHP	S'coach Midland Red
M376MRU	Woodstones Coaches
M396SAB	Hollands
M409PUY	Go Whittle
M421PUY	Go Whittle
M422PUY	Go Whittle
M423PUY	Go Whittle
M425PUY	Go Whittle
M432PUY	Go Whittle
M433PUY	Hollands
M434PUY	Go Whittle
M436PUY	Go Whittle
M437PUY	Go Whittle
M489HBC	Pearces
M501VJO	The Oxford Bus Co
M502VJO	The Oxford Bus Co
M503VJO	The Oxford Bus Co
M504VJO	The Oxford Bus Co
M505VJO	The Oxford Bus Co
M506VJO	The Oxford Bus Co
M507VJO	The Oxford Bus Co
M508VJO	The Oxford Bus Co
M509VJO	The Oxford Bus Co
M510VJO	The Oxford Bus Co
M511VJO	The Oxford Bus Co
M512VJO	The Oxford Bus Co
M513VJO	The Oxford Bus Co
M514VJO	The Oxford Bus Co
M515VJO	The Oxford Bus Co
M516VJO	The Oxford Bus Co
M517VJO	The Oxford Bus Co
M518VJO	The Oxford Bus Co
M519VJO	The Oxford Bus Co
M520VJO	The Oxford Bus Co
M697EDD	Cheltenham & G
M698EDD	Cheltenham & G
M699EDD	Cheltenham & G
M701EDD	Cheltenham & G
M702EDD	Cheltenham & G
M703EDD	Cheltenham & G
M704JDG	Cheltenham & G
M705JDG	Cheltenham & G
M706JDG	Cheltenham & G
M707JDG	Cheltenham & G
M708JDG	Cheltenham & G
M709JDG	Cheltenham & G
M710JDG	Cheltenham & G
M711FMR	Cheltenham & G
M712FMR	Cheltenham & G
M713FMR	Cheltenham & G
M714FMR	Cheltenham & G
M715FMR	Cheltenham & G
M737RCJ	Newbury Coaches
M843EMW	Cheltenham & G
M844EMW	Cheltenham & G
M845EMW	Cheltenham & G
M847HDF	Cheltenham & G
M960VWY	Sargeants
M968RWL	Pearces
MAU145P	S'coach Midland Red
MCJ500P	Rogers of Martley
MCJ900P	Lugg Valley
MCT226X	Astons
MDG193W	Newbury Coaches
MED396P	Barry's Coaches
MEF154J	Rogers of Martley
MFV31T	Nichols
MHB851P	K W Beard Ltd
MHB855P	Lugg Valley
MIW1607	Dudley's Coaches
MIW5785	Wainfleet
MIW5786	Wainfleet
MIW5787	Wainfleet
MIW5788	Wainfleet
MIW5789	Wainfleet
MIW5790	Wainfleet
MIW5791	Wainfleet
MIW5793	Wainfleet
MIW5794	Wainfleet
MIW5795	Wainfleet
MIW5796	Wainfleet
MIW5797	Wainfleet
MIW5798	Wainfleet
MJI1676	Grayline
MJI1677	Grayline
MJI1679	Grayline
MKP181W	Yeomans
MKV87V	S'coach Midland Red
MLH304L	Guide Friday
MNV191P	Guide Friday
MNW133V	Swanbrook
MOI1793	DRM
MOI3512	DRM
MOI4000	DRM
MOI5055	DRM
MOI5633	DRM
MOI9565	DRM
MOU739R	Cheltenham & G
MOU740R	Cheltenham & G
MPT314P	Guide Friday
MRJ8W	The Oxford Bus Co
MRJ9W	The Oxford Bus Co
MSF738P	Yarrington
MUA872P	Cheltenham & G
MUD25W	Chiltern Queens
MUV837X	Primrose Motors
MVK546R	Swanbrook
MVK548R	Swanbrook
MYO486X	Dukes Coaches
N2HCR	Hardings
N3ARJ	Andy James
N3HCR	Hardings
N10JRJ	Johnsons
N26FUY	Go Whittle
N27FUY	Go Whittle
N30ARJ	Andy James
N41MJO	Thames Transit
N42MJO	Thames Transit
N43MJO	Thames Transit
N45MJO	Thames Transit
N46MJO	Thames Transit
N47EJO	Thames Transit
N47MJO	Thames Transit
N48EJO	Thames Transit
N48MJO	Thames Transit
N51KBW	Thames Transit
N52KBW	Thames Transit
N53KBW	Thames Transit
N54KBW	Thames Transit
N56KBW	Thames Transit
N57KBW	Thames Transit
N58KBW	Thames Transit
N59KBW	Thames Transit
N62KBW	Thames Transit
N63KBW	Thames Transit
N64KBW	Thames Transit
N108BHL	Pearces
N114YAB	Redline
N129MBW	Pearces
N156BFC	The Oxford Bus Co
N157BFC	The Oxford Bus Co
N158BFC	The Oxford Bus Co
N171LHU	Tappins
N172LHU	Tappins
N173LHU	Tappins
N174LHU	Tappins
N175LHU	Tappins
N176LHU	Tappins
N177LHU	Tappins
N178LHU	Tappins
N179LHU	Tappins
N180LHU	Tappins
N182CMJ	S'coach Midland Red
N183CMJ	S'coach Midland Red
N201CUD	Thames Transit
N202CUD	Thames Transit
N203CUD	Thames Transit
N204CUD	Thames Transit
N205CUD	Thames Transit
N206CUD	Thames Transit
N206TDU	S'coach Midland Red
N207TDU	S'coach Midland Red
N208TDU	S'coach Midland Red
N211TDU	S'coach Midland Red
N212TDU	S'coach Midland Red
N213TDU	S'coach Midland Red
N214TDU	S'coach Midland Red
N215TDU	S'coach Midland Red
N216TDU	S'coach Midland Red
N301XAB	Midland Red West
N302XAB	Midland Red West
N303XAB	Midland Red West
N304XAB	Midland Red West
N305XAB	Midland Red West
N306XAB	Midland Red West
N307XAB	Midland Red West
N308XAB	Midland Red West
N309XAB	Midland Red West
N310XAB	Midland Red West
N311XAB	Midland Red West
N312XAB	Midland Red West
N313XAB	Midland Red West
N325MFE	Castleways
N341EUY	Midland Red West
N347AVV	S'coach Midland Red
N348AVV	S'coach Midland Red
N349AVV	S'coach Midland Red
N350AVV	S'coach Midland Red
N351AVV	S'coach Midland Red
N352AVV	S'coach Midland Red
N353AVV	S'coach Midland Red
N354AVV	S'coach Midland Red
N355AVV	S'coach Midland Red
N356AVV	S'coach Midland Red
N357AVV	S'coach Midland Red
N358AVV	S'coach Midland Red
N359AVV	S'coach Midland Red
N360AVV	S'coach Midland Red
N361AVV	S'coach Midland Red
N362AVV	S'coach Midland Red
N363AVV	S'coach Midland Red
N364AVV	S'coach Midland Red
N365AVV	S'coach Midland Red
N366AVV	S'coach Midland Red
N367AVV	S'coach Midland Red
N368AVV	S'coach Midland Red
N369AVV	S'coach Midland Red
N370AVV	S'coach Midland Red
N371AVV	S'coach Midland Red
N372AVV	S'coach Midland Red
N383EAK	McLeans
N384EAK	McLeans
N401LDF	Cheltenham & G
N402LDF	Cheltenham & G
N403LDF	Cheltenham & G
N404LDF	Cheltenham & G
N405LDF	Cheltenham & G
N406LDF	Cheltenham & G
N407LDF	Cheltenham & G
N408LDF	Cheltenham & G
N409LDF	Cheltenham & G
N413NRG	The Oxford Bus Co
N414NRG	The Oxford Bus Co
N415NRG	The Oxford Bus Co
N416NRG	The Oxford Bus Co
N521MJO	The Oxford Bus Co
N522MJO	The Oxford Bus Co
N523MJO	The Oxford Bus Co
N524MJO	The Oxford Bus Co
N601FJO	The Oxford Bus Co
N602FJO	The Oxford Bus Co
N603FJO	The Oxford Bus Co
N604FJO	The Oxford Bus Co
N605FJO	The Oxford Bus Co
N606FJO	The Oxford Bus Co
N607FJO	The Oxford Bus Co
N608FJO	The Oxford Bus Co
N609FJO	The Oxford Bus Co
N610FJO	The Oxford Bus Co
N611FJO	The Oxford Bus Co
N612FJO	The Oxford Bus Co
N613FJO	The Oxford Bus Co
N614FJO	The Oxford Bus Co
N615FJO	The Oxford Bus Co
N616FJO	The Oxford Bus Co
N617FJO	The Oxford Bus Co
N618FJO	The Oxford Bus Co
N619FJO	The Oxford Bus Co
N61KBW	Thames Transit
N620FJO	The Oxford Bus Co
N621FJO	The Oxford Bus Co
N622FJO	The Oxford Bus Co
N623FJO	The Oxford Bus Co

N624FJO	The Oxford Bus Co
N649KWL	Pearces
N680RDD	Pulham's
N681AHL	Grayline
N716KAM	Cheltenham & G
N717KAM	Cheltenham & G
N718RDD	Cheltenham & G
N719RDD	Cheltenham & G
N720RDD	Cheltenham & G
N721RDD	Cheltenham & G
N722RDD	Cheltenham & G
N723RDD	Cheltenham & G
N724RDD	Cheltenham & G
N725RDD	Cheltenham & G
N726RDD	Cheltenham & G
N727RDD	Cheltenham & G
N728RDD	Cheltenham & G
N729RDD	Cheltenham & G
N730RDD	Cheltenham & G
N731RDD	Cheltenham & G
N732RDD	Cheltenham & G
N733RDD	Cheltenham & G
N734RDD	Cheltenham & G
N735RDD	Cheltenham & G
N760NAY	Astons
N901PFC	Thames Transit
N902PFC	Thames Transit
N903PFC	Thames Transit
N914DWJ	Astons
N915DWJ	Astons
NAB250P	Hardings
NAD600W	Pulham's
NAK28X	S'coach Midland Red
NAK29X	S'coach Midland Red
NAL53P	Carterton
NCJ800M	Lugg Valley
NDD113W	Carterton
NDE147Y	Cottrell's
NDF857W	Soudley Valley
NDF858W	Soudley Valley
NFB114R	Cheltenham & G
NFB603R	Cheltenham & G
NFH200W	Pulham's
NFH528W	Marchants
NFH530W	Marchants
NHH382W	Cheltenham & G
NHU671R	S'coach Midland Red
NHU672R	S'coach Midland Red
NIA5055	Astons
NIW7756	Kestrel
NJI5235	Teme Valley
NMJ284V	Marchants
NMJ286V	Marchants
NNT522P	Yeomans
NNW119P	K W Beard Ltd
NOC595R	Westward Travel
NOE551R	S'coach Midland Red
NOE553R	S'coach Midland Red
NOE554R	S'coach Midland Red
NOE571R	S'coach Midland Red
NOE577R	S'coach Midland Red
NOE578R	S'coach Midland Red
NOE581R	S'coach Midland Red
NOE582R	S'coach Midland Red
NOE586R	S'coach Midland Red
NOE587R	S'coach Midland Red
NOE589R	S'coach Midland Red
NOE590R	S'coach Midland Red
NOE602R	S'coach Midland Red
NOE603R	S'coach Midland Red
NOE604R	S'coach Midland Red
NOE605R	S'coach Midland Red
NOE606R	S'coach Midland Red
NPA230W	S'coach Midland Red
NRU308M	Grayline
NTC132Y	Cheltenham & G
NUW635Y	The Oxford Bus Co
NUW661Y	The Oxford Bus Co
NUW667Y	The Oxford Bus Co
NUY312T	Lugg Valley
NVJ150M	K W Beard Ltd
NWS289R	Cheltenham & G
NWS903R	H & H Motors
NWT637P	Charlton Services
NWT639P	Charlton Services
NXI918	De Luxe
OAD200P	Soudley Valley
OBD842P	S'coach Midland Red
ODF561	Pulham's
ODM777L	Soudley Valley
ODN601	Go Whittle
ODW459	Alexcars
OFR934T	Primrose Motors
OGL849	Cheney Coaches
OHA436W	Hollands
OHE275X	K W Beard Ltd
OHR190R	Westward Travel
OHV188Y	Bromyard Omnibus
OHV711Y	The Oxford Bus Co
OHV727Y	The Oxford Bus Co
OHV745Y	The Oxford Bus Co
OHV783Y	The Oxford Bus Co
OHY791R	Yarrington
OIB9385	JBC-Malvernian
OIB9386	JBC-Malvernian
OIB9387	JBC-Malvernian
OJD151R	Swanbrook
OJD401R	Worth's
OJI3907	Charlton Services
OJO835M	Chiltern Queens
OKV399W	Smith's
ONH846P	S'coach Midland Red
ORS86R	Bromyard Omnibus
OSF939M	Tappins
OTO543M	Guide Friday
OTO549M	Guide Friday
OTO552M	Guide Friday
OTO571M	Guide Friday
OTO573M	Guide Friday
OTO574M	Guide Friday
OTO582M	Guide Friday
OTO584M	Guide Friday
OTO585M	Guide Friday
OVJ700R	Lugg Valley
OWK83W	Lewis's
OXI9100	Charlton Services
P101NHN	S'coach Midland Red
P102NHN	S'coach Midland Red
P103NHN	S'coach Midland Red
P104NHN	S'coach Midland Red
P105NHN	S'coach Midland Red
P63GHE	Astons
PAB911T	Yeomans
PAD806W	Cottrell's
PBO11Y	Bennetts
PCJ900R	Yeomans
PDF567	Pulham's
PDN873	Cheney Coaches
PEU511R	S'coach Midland Red
PEU515R	Cheltenham & G
PEU516R	S'coach Midland Red
PFO300R	Yeomans
PGR619N	Yeomans
PGU995R	DRM
PHR583R	Lewis's
PHW988S	Cheltenham & G
PHW989S	Cheltenham & G
PIB5773	Yarrington
PIB8109	S'coach Midland Red
PJH582X	Chiltern Queens
PJI7230	Bennetts
PJI7755	K W Beard Ltd
PJI8917	N N Cresswell
PKG108R	JBC-Malvernian
PNW321W	Johnsons
PPH468R	Cheltenham & G
PPJ65W	Chiltern Queens
PTD673S	Westward Travel
PTV596X	Yeomans
PUB13W	Sargeants
PUK621R	S'coach Midland Red
PUK622R	S'coach Midland Red
PUK623R	S'coach Midland Red
PUK624R	S'coach Midland Red
PUK625R	S'coach Midland Red
PUK626R	S'coach Midland Red
PUK627R	S'coach Midland Red
PUK628R	S'coach Midland Red
PUK629R	S'coach Midland Red
PUK656R	Midland Red West
PVJ300M	Lugg Valley
PVV316	Heyfordian
PWL999W	The Oxford Bus Co
Q276UOC	Midland Red West
Q305VAB	JBC-Malvernian
Q553UOC	Midland Red West
RBT172M	Carterton
RCE510	Go Whittle
RDU4	Astons
RDZ4287	Yarrington
REU310S	Cheltenham & G
REU311S	Cheltenham & G
RFC10T	Chiltern Queens
RFC12T	Chiltern Queens
RFC443W	Cheney Coaches
RHG880X	Cheltenham & G
RJI8681	Johnsons
RJI8682	Johnsons

The latest arrivals for the Citylink service operated by the Oxford Bus Company between Oxford and London are Volvo B10Ms with Plaxton Première 320 bodies. Photographed when new, 158, N158BFC, is seen in Oxford and blinds set for the X70 service. *Andrew Jarosz*

Registration	Operator
RJI8683	Johnsons
RJI8684	Johnsons
RJI8685	Johnsons
RJI8686	Johnsons
RJI8687	Johnsons
RJI8688	Johnsons
RJI8689	Johnsons
RJI8690	Johnsons
RL2727	Boomerang Bus Co
RNK327W	Johnsons
RPP514	Go Whittle
RPT293K	De Luxe
RRM385X	Cheltenham & G
RUA451W	Bennetts
RUA452W	Bennetts
RUA457W	Bennetts
RUA458W	Bennetts
RUA460W	Bennetts
RUE353W	Johnsons
SAE752S	Cheltenham & G
SAE753S	S'coach Midland Red
SAE754S	Cheltenham & G
SAE756S	Cheltenham & G
SCN247S	Cheltenham & G
SCN250S	Cheltenham & G
SCN252S	S'coach Midland Red
SCN253S	S'coach Midland Red
SCN255S	Cheltenham & G
SCN256S	Cheltenham & G
SCN264S	Cheltenham & G
SCN265S	S'coach Midland Red
SCN276S	S'coach Midland Red
SCN281S	S'coach Midland Red
SDA566S	Swanbrook
SDA776S	Swanbrook
SDD133R	Soudley Valley
SDR450T	JBC-Malvernian
SDR595	Rogers of Martley
SGF483L	Sargeants
SGR133R	H & H Motors
SGS499W	Boomerang Bus Co
SHH389X	Cheltenham & G
SHH392X	S'coach Midland Red
SIB4458	Swanbrook
SIB6441	Sargeants
SIB8340	Chauffeurs
SIJ4712	Go Whittle
SJI4428	Heyfordian
SJI5861	Heyfordian
SJO871T	Sargeants
SKG406Y	Worth's
SND710X	Cheltenham & G
SNJ592R	The Oxford Bus Co
SNS825W	Cheltenham & G
SOA657S	Midland Red West
SOA658S	Midland Red West
SPV860	Grayline
STO244X	Dudley's Coaches
SVV589W	S'coach Midland Red
SWV804	Catteralls
TAD24W	Dukes Coaches
TAE639S	S'coach Midland Red
TAE641S	Cheltenham & G
TAE642S	Cheltenham & G
TAE644S	Cheltenham & G
TBD172N	Sargeants
TDF103R	Swanbrook
TDM770V	Worth's
THX231S	S'coach Midland Red
THX340S	Swanbrook
THX500S	Swanbrook
THX634S	Catteralls
TIB4921	H & H Motors
TIB6410	Swanbrook
TIB6411	Swanbrook
TJF757	Castleways
TND115X	Catteralls
TND431X	Barry's Coaches
TOF10S	S'coach Midland Red
TOF707S	S'coach Midland Red
TOF708S	S'coach Midland Red
TOF709S	S'coach Midland Red
TPJ287S	Go Whittle
TSV804	Chiltern Queens
TUP432R	Bromyard Omnibus
TVH137X	Bennetts
TVM263	Teme Valley
TWP97V	Dudley's Coaches
TWS906T	Cheltenham & G
TWS913T	Cheltenham & G
TWS914T	Cheltenham & G
TXI6704	Johnsons
TXI6705	Johnsons
TXI6706	Johnsons
TXI6707	Johnsons
TXI6708	Johnsons
TXI6709	Johnsons
TXI6710	Johnsons
UAM207	Springs Tours
UBC464X	Marchants
UDF936	Pulham's
UJI1759	The Oxford Bus Co
UJI1760	The Oxford Bus Co
UJI1761	The Oxford Bus Co
UJI1762	The Oxford Bus Co
UJI1763	The Oxford Bus Co
UJI6312	McLeans
UJV831	Nichols
UKV470R	Guide Friday
UKV473R	Guide Friday
UKV479R	Guide Friday
UKV482R	Guide Friday
ULS670T	Astons
UMR197T	Andy James
UMR198T	Andy James
UOI7274	Dudley's Coaches
URB161S	Westward Travel
URH13R	Cheney Coaches
URH341	Go Whittle
URT682	Cheney Coaches
USU800	Cheney Coaches
UTV215S	Guide Friday
UTV217S	Guide Friday
UVK298T	S'coach Midland Red
UVO125S	Worth's
UWA579Y	Ken Rose
UWP96R	Dudley's Coaches
UWW7X	Cheltenham & G
UWY83X	Andy James
UWY84X	Andy James
VAD141	Pulham's
VAE499T	Cheltenham & G
VAE501T	Cheltenham & G
VAE502T	S'coach Midland Red
VAE507T	Cheltenham & G
VBW581	Chiltern Queens
VCU301T	S'coach Midland Red
VCU304T	S'coach Midland Red
VCU310T	S'coach Midland Red
VDF365	Pulham's
VDG700X	Pulham's
VEN416L	Lugg Valley
VEU231T	Cheltenham & G
VGJ317R	Charlton Services
VJA660S	Yarranton
VJO201X	The Oxford Bus Co
VJO202X	The Oxford Bus Co
VJO203X	The Oxford Bus Co
VJO204X	The Oxford Bus Co
VJO205X	The Oxford Bus Co
VJO206X	The Oxford Bus Co
VOD596S	Cheltenham & G
VOD597S	Cheltenham & G
VOI5888	Hardings
VPF287S	Westward Travel
VPF296S	The Oxford Bus Co
VPF742	Cheney Coaches
VPP958S	Charlton Services
VPT965R	Rogers of Martley
VRY610X	Soudley Valley
VSF438	Heyfordian
VTV170S	S'coach Midland Red
VUR896W	Ken Rose
VWK8S	Yeomans
VWX350X	Westward Travel
VYL851S	K W Beard Ltd
WAO397Y	Cheltenham & G
WAS765V	S'coach Midland Red
WCU823T	Westward Travel
WDA994T	S'coach Midland Red
WDD17X	K W Beard Ltd
WDF946	Pulham's
WDF998X	Marchants
WDF999X	Marchants
WJM815T	Springs Tours
WJY759	Guide Friday
WLT713	Cheltenham & G
WOC722T	Midland Red West
WOC723T	Midland Red West
WSU293	S'coach Midland Red
WTL642	Go Whittle
WUD815T	Chiltern Queens
WUG143S	Yeomans
WVJ300T	Yarrington
WVJ530T	Yeomans
WVU152S	Go Whittle
WWL207X	The Oxford Bus Co
WWL208X	The Oxford Bus Co
WWL209X	The Oxford Bus Co
WWL210X	The Oxford Bus Co
WWL211X	The Oxford Bus Co
WWL212X	The Oxford Bus Co
WWL537T	Grayline
WWR417S	Guide Friday
WWR418S	Guide Friday
WWR419S	Guide Friday
WWR420S	Guide Friday
WWY125S	Bennetts
WXI6274	Charlton Services
WYV47T	Swanbrook
XAD174X	Bennetts
XAK457T	Swanbrook
XAK902T	Marchants
XCJ750T	N N Cresswell
XCT550	Heyfordian
XDG614	Pulham's
XDV602S	Cheltenham & G
XDV606S	Cheltenham & G
XGR728R	S'coach Midland Red
XKH455	Go Whittle
XKO54A	Guide Friday
XKV488S	Worth's
XNK201X	Bromyard Omnibus
XNN664S	Guide Friday
XOI1908	Primrose Motors
XOV743T	Midland Red West
XOV744T	Midland Red West
XOV746T	Midland Red West
XOV749T	Midland Red West
XOV752T	Midland Red West
XOV753T	S'coach Midland Red
XOV754T	S'coach Midland Red
XOV755T	S'coach Midland Red
XOV756T	S'coach Midland Red
XOV758T	Midland Red West
XOV760T	S'coach Midland Red
XSK144	Worth's
XSU912	Carterton
XTP287L	Thames Transit
XVC230X	De Luxe
XXI8563	Chauffeurs
XXN665S	Guide Friday
YAY537	Heyfordian
YBO16T	S'coach Midland Red
YBO18T	S'coach Midland Red
YCT502	Yeomans
YCV834	Yeomans
YDD109S	Marchants
YEH182X	De Luxe
YEU446V	S'coach Midland Red
YFB972V	Cheltenham & G
YFB973V	Cheltenham & G
YFC18V	Chiltern Queens
YFR491R	Andy James
YFU846	Go Whittle
YJV806	Cheltenham & G
YLW894X	Yeomans
YLW897X	Primrose Motors
YOI298	Yeomans
YOI2747	Yeomans
YPL72T	K W Beard Ltd
YSD340L	De Luxe
YSU975	Cheney Coaches
YTD384N	Carterton
YUE338	Tappins
YXI2755	Yeomans

ISBN 1 897990 18 9
Published by *British Bus Publishing*
The Vyne, 16 St Margarets Drive, Wellington,
Telford, Shropshire, TF1 3PH
Fax/Orderline (+44) (0) 1952 255669

Printed by Graphics & Print
Unit A13, Stafford Park 15
Telford, Shropshire, TF3 3BB